"This edited volume, drawing on subjects such as philosophy, art history, and pedagogy, makes an extraordinarily original and significant contribution to our understanding of the current post-climate era. By reimagining sustainability and addressing the troubling future of rising global temperatures and social injustice, it explores the intersection of art, education, and philosophy across all chapters. This fusion inspires creativity, not only in the world at large but also in how universities can foster a hopeful and inclusive community, nurturing a future rooted in belonging and shared purpose."

Peter Dobers, *Professor of Business Studies, Head of Department of Journalism, Södertörn University, Sweden*

"Säfström and Loughran reveal a world which we are invited to navigate with even more freedom than we can muster. They do so through the diverse spheres of a sophistic(al) heterodoxy—indeed what they call a 'patchwork ontology', which allows us to move beyond the strictures that have become the mainstay of our current polity. Being in the world anew is what, one hopes, would allow us to survive the destructive legalities by which human agency became a reified mirage. This is a fascinating book which does not simply challenge our certainties, but confirms, yet again, that the only way to make sense of and prevail in such a world, is to embrace form's pedagogical possibilities."

John Baldacchino, *Professor of Art and Education, The University of Wisconsin-Madison, USA*

"An extraordinarily fine collection of authors carry the world events of pedagogy into the realm of thinking otherwise to invite the risk of beauty in our time of ecological crisis. Each chapter is a gem of engagement and provides a treasure trove of educational experiments crafted from the ethics of thinking again. On offer is a new vocabulary as openings to imagination, hospitality, and involvement, with the request to analyze one's own situation within the learning world of others. With the medium for philosophical thought and pedagogical responsiveness, readers enter movements of wonder with fresh thinking, surprising metaphors, and views of life that welcome the coming education."

Deborah P. Britzman, *author of* When History Returns: Psychoanalytic Quests for Humane Learning

EVENTS OF ART AND EDUCATION IN POST-CLIMATE TIMES

This book concerns the urgency of thinking and acting in response to climate change through art and education. While both fields are often connected through disciplinary dialogues, climate change prompts a greater need to unite artists and educators around common environmental problems and goals.

By staging transcritical engagements, this book draws out common and uncommon disciplinary perspectives that can generate new ways of thinking, living, and doing in the Anthropocene. Ideas around courage, resilience, life, and death emerge. An expression of active, non-violent resistance to the ongoing destruction of our planet, this book supports imaginative action, popular sovereignty, and the courage to live well within the challenges of our era. Engaging artists' and educators' questions, it maps significant differences and potential intersections for further enquiry.

Events of Art and Education in Post-climate Times will be helpful for students studying art, education, environment, sustainability, and climate change. It will also interest researchers, practising artists, and teachers in these disciplines by being at the forefront of current discussions in both fields.

Carl Anders Säfström is a professor of educational research and director of the Centre for Public Education and Pedagogy at Maynooth University, Ireland, as well as a Zen meditation practitioner and an aspiring poet. His research currently focuses on the tradition of education as a practice established by the Sophists, a critique of the Platonic–Aristotelian orthodoxy dominating educational thought, and the violence, systemic as well as symbolic, following in education from such orthodoxy.

Glenn Loughran is an artist and educator at the Technological University Dublin. He is co-founder and programme director of the archipelagic MA in Art and Environment, set up in 2020, and delivered across the Islands of West Cork. His research focuses on artistic research, socially engaged art, critical pedagogy, and island studies. He is a member of the Working Group on Artistic Research at the European League of Institutes of the Arts (ELIA).

ROUTLEDGE STUDIES IN SUSTAINABILITY

Critical Sustainability Sciences
Intercultural and Emancipatory Perspectives
Edited by Stephan Rist, Patrick Bottazzi and Johanna Jacobi

Transdisciplinary Research, Sustainability, and Social Transformation
Governance and Knowledge Co-Production
Tom Dedeurwaerdere

Digital Technologies for Sustainable Futures
Promises and Pitfalls
Edited by Chiara Certomà, Fabio Iapaolo, and Federico Martellozzo

Rethinking the Concept of Waste and Mass Consumption
Preserving Resources through Reuse, Repair and Recycling
Richard Waite

Mediating Sustainability in the Consumer Society
Edited by Astrid Skjerven, Lisbeth Løvbak Berg, Liv Merete Nielsen, and Dagny Stuedahl

Smart Cities to Smart Societies
Moving Beyond Technology
Edited by Esmat Zaidan, Imad Antoine Ibrahim, and Elie Azar

Events of Art and Education in Post-climate Times
Carl Anders Säfström and Glenn Loughran

For more information on this series, please visit: www.routledge.com/Routledge-Studies-in-Sustainability/book-series/RSSTY

EVENTS OF ART AND EDUCATION IN POST-CLIMATE TIMES

Edited by Carl Anders Säfström and Glenn Loughran

Routledge
Taylor & Francis Group
LONDON AND NEW YORK

Designed cover image: Glenn Loughran

First published 2026
by Routledge
4 Park Square, Milton Park, Abingdon, Oxon OX14 4RN

and by Routledge
605 Third Avenue, New York, NY 10158

Routledge is an imprint of the Taylor & Francis Group, an informa business

British Library Cataloguing-in-Publication Data
A catalogue record for this book is available from the British Library

ISBN: 978-1-032-59865-9 (hbk)
ISBN: 978-1-032-58683-0 (pbk)
ISBN: 978-1-003-45663-6 (ebk)

DOI: 10.4324/9781003456636

Typeset in Times New Roman
by Taylor & Francis Books

CONTENTS

CONTRIBUTORS

Juliette Clara Bertoldo is a researcher, cross-disciplinary thinker, and art lover. Her journey began with professional dance training and extensive work in visual media, devoting over ten years to experimenting with aesthetic and embodied modes of creation. Alongside her artistic practice, she completed an MA in Educational Studies at the University of Glasgow in 2018 and recently defended her PhD in the Department of Education at Maynooth University. Her work explores the educational value of death, art, and aesthetics in a time of planetary unrest and thinks predominantly with critical posthumanism, queer death studies, and extinction studies. She has published a number of articles on the relevance of these interrelated issues to education in the Anthropocene, the latest to date in *Research in Arts and Education* and *Philosophy of Education Society.*

Gert Biesta (www.gertbiesta.com) is professor of public education in the Centre for Public Education and Pedagogy at Maynooth University, Ireland, and has further academic affiliations in Scotland and Norway. He writes about the theory of education and the philosophy of social and educational research, with a particular interest in teaching, teachers, curriculum, policy and democracy. He is a member of the Education Council of the Netherlands, the constitutional advisory body of the Dutch government and parliament. Recent books include *The New Publicness of Education: Democratic Possibilities after the Critique of Neo-liberalism*, co-edited with Carl Anders Säfström (2023), and *World-Centred Education: A View for the Present* (2022). He is deeply concerned about the ways in which the global education measurement industry continues to undermine education as a public concern, and sees his scholarship as an act of resistance against the trivialisation of education and the work of teachers. He takes the fact that his work has been translated into over twenty languages as an indication that these concerns are shared across the globe.

Barbara Knežević is an artist and lecturer at Technological University Dublin. She makes sculpture, installation and moving image artworks that explicate the historical, material, semiotic and political functions of materials and objects. Recently, her practice has become attentive to how sculptures are instrumentalised and affect the politics of ethnicity and personal and national histories, historically and in contemporary contexts. Her work is included in public and private collections internationally and nationally such as the Irish Museum of Modern Art and the Arts Council of Ireland. Knežević's large-scale sculpture and film project titled *The Iron Gates* (Gvozdene Kapije), focusing on the archaeological site of Lepenski Vir, Đjerdap hydroelectric dam, and the region surrounding the Iron Gates in Serbia, opened at Solstice Arts Centre in March 2025 and will tour several venues in Ireland and Europe in 2026.

Glenn Loughran is an artist and educator at the Technological University Dublin. He is co-founder and programme director of the archipelagic MA in Art and Environment, set up in 2020, and delivered across the Islands of West Cork. His research focuses on artistic research, socially engaged art, critical pedagogy, and island studies. Recent publications include *What Is an Island: Towards a Patchwork Pedagogy for a World Archipelago* (2023). He has exhibited at the Cooper Gallery, Dundee, Helsinki Research Pavilion, Venice, Medialab Prado, Madrid, Lithuania Biennale, Kaunas, the Irish Museum of Modern Art, Dublin, the Hugh Lane Gallery, Dublin. His research has been presented at the Creative Time Summit, Venice Biennale, Centre Pompidou Paris, ELIA Biennale and published as book chapters by Routledge, Columbia University Press, and Sternberg Press. He is a member of the Working Group on Artistic Research at the European League of Institutes of the Arts (ELIA).

Paul O'Keeffe is head of education and research at Airfield Estate, a working farm, research and education facility, and visitor attraction that promotes and provides sustainable agriculture and food education in Dublin. He is an expert in the role that education can play in addressing global issues and has extensive experience of managing education and development programmes in Europe, Africa, and the Middle East. He has researched and published extensively on enabling education in refugee and crisis contexts, and his current research focuses on refugee-led development and sustainable agricultural education in Kakuma refugee camp in Kenya, where he has worked since 2017.

Leif Östman is professor, director, and one of the founders of the discipline of didactics at Uppsala University. He is leading the international research group TePlab–The Laboratory for Teaching Practices, where the focus is on teaching as well as facilitation of non-formal learning, and the consequences it has for participants' learning, actions and self-cultivation. Leif has been leading numerous projects and supervised over 20 doctoral students, mainly in science and sustainability education. He was the director of the Swedish research school on environmental and sustainability education and has been working together with teachers, teacher trainers, policy makers, etc. in Mongolia and other countries to develop participatory and locally relevant sustainability teaching. The experiences from this work and collaboration with TePlab colleagues resulted in the development of a specific form for cooperation between researchers and teachers: lesson design workshops. The cooperation is focused on co-creating lesson plans, teaching methods, and teaching materials which can be widely shared through tailor-made digital platforms. Besides being heavily involved in empirical analyses, he has also developed theoretical models and analytical methods based on pragmatist philosophy and the later works of Ludwig Wittgenstein as well as results of and frictions experienced in empirical studies. Leif has a background as a science teacher in a lower secondary school and is since 1987 involved in teacher training and in-service training. He is also devoted to handicraft, nowadays as a hobby.

Carl Anders Säfström is a professor of educational research and director of the Centre for Public Education and Pedagogy at Maynooth University, Ireland, as well as a Zen meditation practitioner and an aspiring poet. He is also an affiliated professor of education at Malmö University, Sweden. His previous positions include a Baltic Sea Foundation professor of education at Södertörn University, Sweden, professor of education and dean of teacher education at Mälardalen University, Sweden and a period as president of the Nordic Educational Research Association (NERA). His research currently focuses on the tradition of education as a practice established by the Sophists, a critique of the Platonic–Aristotelian orthodoxy dominating educational thought, and the violence, systemic as well as symbolic following in education from such orthodoxy. He co-edited his latest book for Routledge with Gert Biesta (2023), *The New Publicness of Education: Democratic Possibilities after the Critique of Neo-liberalism*. Springer published his latest single-authored book, *Education for Everyday Life: A Sophistical Practice of Teaching*, in 2023. In his 2021 book for Routledge, *A Pedagogy of Equality in a Time of Unrest: Strategies for an Ambiguous Future*, he started to move more explicitly towards urgent themes in today's world in his writing.

Timothy Stott is associate professor in modern and contemporary art history and head of the Department of History of Art and Architecture at Trinity College Dublin. His research focuses on the history of environmental art and design and, most recently, the visual culture and design of climate science. Recent publications include the monograph *Buckminster Fuller's World Game and Its Legacy* (2021), and the edited collection (with Johanna Gosse) *Nervous Systems: Art, Systems, and Politics since the 1960s* (2022). He is a member of the Trinity Centre for Environmental Humanities.

Sharon Todd is professor of education and member of the Centre for Public Education and Pedagogy at Maynooth University, Ireland. She has published widely in the areas of relationality, social justice and ethics and is currently engaged in making connections between the climate emergency, art practice, and political aesthetics in education. She is author of *The Touch of the Present: Educational Encounters, Aesthetics, and the Politics of the Senses* (SUNY Press, 2023), *Toward an Imperfect Education: Facing Humanity, Rethinking Cosmopolitanism* (Paradigm, 2009), and *Learning from the Other: Levinas, Psychoanalysis and Ethical Possibilities in Education* (SUNY, 2003). Most importantly, she is a practising Zen Buddhist and abstract painter.

Ellen Vandenplas holds a master's degree in educational sciences. Since 2019 she is affiliated as a doctoral researcher at Ghent University's Department of Political Sciences. Her PhD 'calling the existential into existence in climate change education' focus how teachers can 'see' and 'handle' existential concerns of students in climate change education. In this ongoing doctoral study she first studied students expressions of the existential in climate change education to further conceptualise the existential dimension in climate change education. In this study she also presents theoretical and analytical approaches to examine the didactic work of the teacher to elicit and handle the existential in climate change education. Since 2017, she also supports courses and faculties at Ghent University integrating sustainability in their educational programs. Previously she worked for Ecocampus—the Sustainable Higher Education Program of the Flemish government—where she supported universities and colleges to integrate sustainable development in their education, research and social services. She started her career as a project coordinator for the Center for Nature and Environmental Education.

Katrien Van Poeck is a research professor at Ghent University (Belgium), she coordinates the Centre for Sustainability Studies' research programme on sustainability education. Focusing on the role of teaching and learning in the pursuit of building a more sustainable world, her research has been largely inspired by pragmatist didactic theory. She leads and supervises research projects on facilitating (informal) learning in the context of sustainability transitions, sustainability teaching in higher education, climate change education, and working with real-world sustainability problems in teaching and learning practices. Katrien has a special interest in the relation between sustainability education research and practice. She is strongly engaged in initiatives to foster the co-creation of sustainability education practices in collaborative settings that gather researchers and practitioners, such as Lesson Design Workshops. She is also a member and co-founder of the international research collective TePlab–The Laboratory for Teaching Practices, coordinated at Uppsala University (Sweden). Before joining the Centre for Sustainability Studies in 2015,

Katrien worked in NGOs and youth work organisations (1999–2002), served as a policy advisor for environmental and sustainability education at the Flemish government (2002–2015), and was a voluntary scientific staff member at Leuven University's Laboratory for Education and Society (2009–2015).

Fiona Woods is an artist and educator who works with aesthetics and critical spatial practice, often in a co-productive capacity with others. Her practice combines artistic research, critical inquiry and aesthetic work, and she employs social, public and institutional circuits to explore ideas of what we have 'in common'. Her PhD, 'A Laboratory of Common Interest: Contesting the Economisation of Space in Limerick City through the Practice of Aesthetic Work' (2021) was completed through the Technological University Dublin. She is a lecturer at Limerick School of Art & Design, Technological University of the Shannon, Ireland.

PREFACE

The events leading to this book started as a conversation on a small plane from Finland to Sweden in 2021. Professor Peter Dobers, then dean at Södertörn University, Stockholm, asked me if I would be willing to be part of a keynote panel and to take charge of a track on Education at the 28th Annual Conference of the International Sustainable Development Research Society on Sustainable Development and Courage: Culture, Art and Human Rights, to be arranged in June 2022. After careful consideration, I contacted my colleague and good friend, Dr Glenn Loughran, artist and educator at TU Dublin and program chair of the MA Art and Environment in West Cork. I invited him to work with me on this project. We focused our session on *resilient subjectivities in art and education*, which produced a lively discussion with many key stakeholders in the room. Having worked well together, we moved the project forward, expanding our in-depth conversations around the reach of art and education in climate change. At that point, I was involved in a network project, 'The Public Role of Education in Democratic Sustainability Transitions' (dnr. 2019-04819), funded by the Swedish Research Council, which gave me access to some funding. This network funded a follow-up symposium, 'Art and Education in an Uncertain World', which we arranged at the Centre for Public Education and Pedagogy (CPEP) at Maynooth University in November 2022, a research centre for which I am currently the director. Contributions to the symposium were a broad mix of disciplines and positions that eventually led to this book. So, thank you, Peter, for the invitation, and the Swedish Research Council for the funding. Thanks also to the authors who have devoted their time and energy to the above events and those contributing to this anthology. We believe that we have written an important book together, and we hope it will reach many readers who, like us, are concerned about the state of the world and are committed to speaking up and acting accordingly.

Carl Anders Säfström and Glenn Loughran
Dublin, Stockholm, Maynooth, and Kabusa

INTRODUCTION

Being in the World Anew

Carl Anders Säfström and Glenn Loughran

In 2012, John W. Hessler published the book *Seeing the World Anew: The Radical Vision of Martin Waldseemuller's 1509 & 1516 World Maps.* Telling the story of how the world was successively mapped to secure a new worldview, a world that could be controlled, explored, and colonised. Since then, we have lived in the mapping of our external and inner worlds, of our social, political, emotional, and economic lives. Today, it is as if mapping has exhausted itself as we enter into unknown, uncertain territories after centuries of certainty. Knowing the conditions of this certainty, we are faced with an unprecedented climate event forcing us to shift our perspective about this planet and ourselves. The shifting of perspectives has motivated this publication in several ways. Firstly, we need to consider how climate change produces new perspectives on our increasingly vulnerable place in the world; second, we need to consider how these shifts impact our understanding of education and art. Thirdly, we must discuss the importance of creativity and creation in this new situation since that seems to be precisely what is excluded by the abstraction of maps.

British anthropologist Tim Ingold has made a particular case for the significance of creation and creativity within this shift, arguing that creation 'holds the promise of a world to come, of the continuity of life, and the possibility of renewal' (Ingold, 2024, p. 4). Tracing the origins of the concept of creation, Ingold first draws our attention to the original meaning of the word *nature,* which derives from the Latin verb *nascere* ('to be born'). Following this, he points to Roman poet Lucretius (*c.*99 BC–*c.*55 BC), whose didactic poem *De Rerum Natura* describes the event of creation in the feminine, as the 'creatress' (Ingold, 2024, p. 6).

DOI: 10.4324/9781003456636-1

Importantly for this publication, it is in *De Rerum Natura* that creation is first articulated as an event, as clinamen.

> When atoms move straight down through the void by their own weight, they deflect a bit in space at a quite uncertain time and in uncertain places, just enough that you could say that their motion has changed. But if they were not in the habit of swerving, they would all fall straight down through the depths of the void, like drops of rain, and no collision would occur, nor would any blow be produced among the atoms. In that case, nature would never have produced anything.
>
> *(Lucretius, 1994, p. 66)*

The 'evental swerve' at the centre of *De Rerum Natura* is also the hinge upon which Stephen Greenblatt's historical reading of modernity is narrated, arguing that the atomism at the heart of *De Rerum Natura* formed a rupture in the medieval period, pointing us towards modernity and away from an extended medieval period (Greenblatt, 2011). As we find ourselves again at the beginning of a new epoch, the relationship between the rupture and renewal, event and change are becoming increasingly urgent. As Gaia Vince has suggested, we now live in a 'post-climate world'. Climate change has *already* happened. We are *already* in it; there is no going back.

And yet, the event of climate change is still extremely challenging and difficult to grasp and respond to (Vince, 2024). Navigating these challenges, Iddo Tavory and Robin Wagner-Pacifici identify 'ambiguity' as a unique characteristic of the event of climate change (Tavory and Wagner-Pacifici, 2022). That is, contrary to our common understanding of events, climate change is not identifiable as a specific event located at a particular time and place. It does not have a single tipping point. At stake within this analysis is an attempt to identify a specific set of research problems arguing that, if an event is always defined by the relationship between 'rupture and form', then it is only through a reflection on post-evental disciplinary formation that we can begin to address the ambiguity of climate change and to come to terms with our post-climate change condition. They write 'paying attention to the demonstrative relationship between rupture and temporality allows us to bring the confusions and elisions of climate change narratives into sharper relief' (Tavory and Wagner-Pacifici, 2022, p. 10). In a similar fashion, this publication began as a series of research events and symposia to look at the event of climate change through the lens of education and art to see what disciplinary forms have begun to emerge and to set them in dialogue with each other.

Initiating this research in June 2022, we organised a panel on resilience in art and education at 'Sustainable Development and Courage: Culture,

Art and Human Rights', the 28th Annual Conference of the International Sustainable Development Research Society. Titled 'Beyond the Resilient Subject: Art, Education, and Imagination', this panel asked how art and education could provide alternative spaces of imaginative action to contemporary discourses on resilience and the political subjectivity they promote. While resilience is vital to face the growing difficulties of the Anthropocene, our concern in this panel was that current discourses coupling an ecological subject with an economic subject replace the political tools of collective bargaining and resistance with those of individual survival and adaptation to the market economy and ecological crisis. More specifically, Irish educational policy promotes resilience as a requirement for wellbeing, and in Sweden, the self-investing resilient individual is upheld as an ideal citizen in the neo-conservative turn in educational policies. Within this context, the prevalent images are of those who survive rather than flourish. Following this, the panel proposed a renewed role for the imagination in the art and education of the Anthropocene to promote courage, collectively and human flourishing rather than individual resilience and survival. Through these discussions, we explored *active*, non-violent resistance to the ongoing destruction of our planet and the popular sovereignty needed to live well within it.

This inquiry led to a follow-up symposium at Maynooth University in November 2022, where we explored contemporary shifts in educational thought, from adaptive, binary modes of resilience to entangled and diffracted relations of becoming. Titled 'Art and Education in an Uncertain World', this symposium concerned a greater need to unite artists and educators around common problems and goals in the Anthropocene. Staging a transcritical discussion between art and education, this event addressed key environmental questions and issues methodologically; that is, by looking at environmental challenges through the lens of educational practice and artistic practice, we aimed to draw out common and uncommon disciplinary perspectives that can generate new ways of thinking, doing and living in the Anthropocene. Reflecting on the emergence of new educational forms within these discursive events, we became increasingly attuned to the capacity of fabric philosophies, weaving, stitching, knotting, and patchworking to represent the complexity of creative entanglements in our world (Paavolainen, Teemu, 2017). The following section briefly sketches a patchwork pedagogy for our post-climate times.

Patchworking Pedagogy

This book hinges on the idea that the event of the Anthropocene ruptured the educational worldview inherited from Platonic–Aristotelian rationality, which led to the rationalisation and instrumentalisation of education in modernity. In the ruins of the modern worldview, we must consider new education experiences as critically open, on the move, and disciplined about this uncertain future. Historically, the binary tension between philosophy and education, humans and plants, state and the individual, eternal truth and conditional truth, and the esoteric and the professional are all a consequence of the Platonic–Aristotelian orthodoxy in education. There is, however, throughout intellectual history, sometimes suppressed and sometimes parallel, heterodoxic educational thought and practice. Education in the current publication belongs to the latter. Highlighting practices that explore alternative frameworks for imaginative action in the climate crisis, we aim to draw out their 'anachronistic possibility' within the policing orders of the educational present (Ryther, 2009). Since Plato's *Republic*, the dominant educational theory locates the subject in education within a pattern linking man and state. That is, since Plato's education is primarily understood in terms of *socialisation*, the process through which someone becomes a member of the society in which he or she lives. To be socialised means to share the values and norms of the group to which one belongs so that one can both be recognisable as a member of that group as well as develop one's abilities and talents within such a frame, to become a happy and well-functioning citizen (Durkheim, 1956). In addition, modern education also includes the idea of *qualification*, that we, through education, are qualified to take our part in the social division of labour, that we are competent to build a house, treat a patient, or fly a plane. In addition to being a recognisable part of the society in which one lives and participates through a societal function, one also realises oneself through the subjectification of education (Biesta, 2017).

One is no longer just anyone but someone able to distinguish one's individuality and existential uniqueness. Those three aims or functions of education are necessarily entangled and mutually interdependent. The problem, though, is affecting how those triple functions of education operate in a social setting, which is where education as such is located. We need education to live well with other people every day. In that case, the meaning of those functions will operate quite differently than if the aim and goal of education is for the individual to perfect herself in the image of a perfect state. If education is already understood from the state's point of view, then its sole function, the overriding function of education, is to perfect the state. Werner Jaeger points this out when he describes the educational function of Plato's *The Republic* in forming a perfect state or nation:

> The perfect man can be shaped only within the perfect state and vice versa. That is the ground for the universal correspondence between

> man and state's inner structure, for the resemblance between both patterns. To construct such a state, we must discover how to make such men.
>
> *(Jaeger, 1986 [1943], p. 259)*

In a Platonic worldview, education's role is to construct complementary patterns in which self-realisation will ultimately strengthen and perfect the state. In the Platonic worldview, the will of the state to be a dominant power reproducing itself over time means that resilience is an expression of how well such power can endure over time and is, as such, profoundly political. The resilient subject in Platonic-informed educational *theory* appears at a particular *location* and for a specific reason: to perfect the state by perfecting oneself. From such an understanding of education, it also follows that if the state is in trouble, education can solve it by better educating 'the patterns' through which the state and the individual are perfected. The idea that education can achieve perfection is also why policies for education worldwide tend to see education as an instrument for solving any problem arising; if so, war or climate emergency: everything can be better with better education (Säfström, 2020). If there is a serious problem, education is the answer. Resilience in education demands the individual to strengthen oneself and the pattern through which the state, any state, is perfected as such.

Following the Platonic–Aristotelian line of thought, Immanuel Kant (2015) sees education as the means to fulfil what imperfect nature cannot achieve by itself; education and teaching are to perfect God's creation (Cassin, 2014). For Kant, as for Aristotle, the imperfectness of nature and humans raises the problem of radical evil. The imperfectness of humanity, its inner radical evilness, and its inner nature need to be suppressed, controlled, and civilised through education and teaching for humanity to fulfil God's plan, which can only be perfection. According to Kant, the child must be turned from 'radical evil and self-love' to ethical norms and rules (Cassin, 2014). Nature is separated from humans, and moral perfection is achieved within education in which nature is to be controlled as it gives rise to radical evil. In this line of thought, resilience in education measures how well one manages to control nature within oneself and 'outside' oneself to perfect both nature's work and the state in the name of God.

Platonic/Aristotelian philosophy is immensely influential in establishing the very parameters through which 'our Western way of life' emerges, from the way cities and states are organised to the foundations of science and culture, the invention of education and teaching as means for perfecting the state, and the conditions for social life as we know it, aimed to *dominate* education as a practice of ordinary life. The Plato/Aristotelian universe was to dominate the Sophists' practice-oriented world in establishing theoretical science 'to be true, the 'highest' kind of culture and replace or *dominate* the current educational discipline' as established by the Sophists (Jaeger, 1986 [1943], p. 318; our emphasis). Such domination established a Platonic/Aristotelian *orthodoxy* in which education is all about man's domination over nature and the state's domination over the subject in the name of perfection. Sophistic educational practice, on the other hand, did not build on science as the first philosophy, nor did it separate so harshly from nature, but instead considered how to live well together with a multiplicity of others in the practical and concrete conditions of the 'here and now' (Jaeger, 1965 [1939], 1986 [1943]).

Sophistic education, therefore, opens a *heterodoxic* tradition of thought within intellectual history, in which education is a practice which concerns itself with the first question of such practice: how to live well with oneself as well as a multiplicity of other beings within an imperfect world and nature? Heterodoxic thought is expressed through a non-foundational worldview in which the attention is directed towards life's polyvalence, changing energies and flows, its ambiguities and movements, rather than fixating essences and the absolute and eternal as conditions for engaging with people, animals and plants (Consigny, 2001). The ambition of a heterodoxic practice is not to fixate on this as that, but rather to move in concert with one's surroundings, if so people, animals or plants. The Sophists were not so much interested in 'essence' as the conditions under which we live and realised that those conditions shift over time and space and that we, therefore, need to continuously find ways of relating anew to live well with ourselves as well as others in every concrete environment, on which we depend for our survival. Sophistic education is not about perfecting oneself and the state; there is no 'radical evilness' of nature but concerns living well with others 'here and now' in an imperfect world and nature and a democratic city-state founded on equality.

The Sophists were democrats, embracing a radical plurality of world views as the reason and motivation for the democratic organisation of the city-state. Democracy was the social and political organisation of the equality they found in nature (Jaeger, 1965 [1939], p. 324). The Sophists walked from city-state to city-state to teach and experience directly and concretely how life differed in those city-states and, therefore, developed a teaching directed to how to be responsive and how to 'move' with others different from oneself in those different circumstances. They were teachers and not philosophers. Conversely, the philosopher says Plato had to search for truth for the reason of truth alone and that only certain people, the philosopher himself, under certain circumstances, were able to find this truth, from which power flows to rule all others—for the Sophists, the philosophers, nor the aristocratic man had such a privileged position. Therefore, all and everyone could be taught to rule. As Susan Jarratt (1998) points out, the consequence of Sophists being professional teachers, that they would teach anyone that paid them, *de facto* moved education away from the Platonic and Aristotelian idea that *arete'*, the way we embody and carry our culture, could only be inherited by blood, to be instead something that *anyone* could achieve by study and being taught. That made them the primary target for Plato's violent detestation (Jaeger, 1965 [1939]) and Aristoteles's branding of them, together with 'women, children and slaves' (Cassin, 2014, 6) as *plants,* as not human.

While Plato may have wrongly accused the Sophists of bringing education into the marketplace, the late twentieth and early twenty-first centuries saw the further expansion of marketplace ideologies in education, ironically, accelerated by the Platonic/Aristotelian domination in educational thought. In this split, the separation between man and nature, education and environment intensified, with man's dominance over nature reaching its apex in late modernity. Through the mid to late fifties, education policies were increasingly aligned with economic policies developed by the Chicago School of Economics (Bouchard, 1998). Central to this shift was a theory of human capital based on the financial pillars of market flexibility, utility maximisation and resilience (Bouchard, 1998). Bolstering the universalising project of modernisation, human capital theory advanced the idea that human subjects could be perceived as commodity forms through the lens of rational choice theory. In the second half of the twentieth century, education was seen as the training ground for economic subjects, with self-investment, skills acquisition and knowledge accumulation being weaponised in response to a radically unpredictable economic world (Bouchard, 1998). Extending the 'self-perfection for the State' in the ancient world, human capital subjects became 'educational Rambos' defined by 'competitive individualism', a-political positionality and 'resilient subjectivities' (Baptiste, 2001, p. 197).

While the theory of human capital has dominated policy frameworks in most western states, it is crucial to acknowledge how the alignment of human capital within all modernising states was uneven and inconsistent (O'Sullivan, 2005). Importantly, in countries like Ireland, where modernity came late, the theory of human capital also arrived late. In his broad historical analysis, *Cultural Politics and Irish Education since the 1950*, Denis O'Sullivan (2005) has argued that the concept of human capital did not take hold in Ireland until the late eighties and early nineteen-nineties, emerging only through a long period of inter-textual struggle between the church and state (Ibid). For O'Sullivan, this period is perceived as an ongoing struggle between a 'theocentric paradigm' grounded in the values of the church and state and the 'mercantile paradigm' motivated by ideologies of the free market (O'Sullivan, 2005, p. 225).

To look at educational modes of exchange through the lens of modernity will always lead to interpretations of resilience determined by dominant economic value systems. At the beginning of the twenty-first century, the United Nations Charter highlighted the importance of resilience for contemporary societies, describing it as 'the capacity of a system, community or society to resist or change in order that it may obtain an acceptable level of functioning' (United Nations, 1945). Resilience theorist Julian Reid has argued that the contemporary concept of resilience did not only develop out of classical neo-liberal subjectivity, defined here as human capital but also out of the oppositional discourse of sustainable development (Reid, 2019). Within this critique, Reid highlights the tendency in popular discourses on resilience to reduce the concepts of imagination and resilience to survivability alone. He writes, 'Imagination can either contribute to the survival strategies with which human beings attempt to care for themselves in the face of ordeals and traumas or it can, more ambitiously, seek to create an image of the self, existing free from the possibility and necessity of a life of endless trauma and struggle' (Reid, 2019, p. 37).

Recent shifts in Anthropocene thinking have begun to challenge such ideological systems, arguing that modern governance's top-down, command-and-control structures are no longer viable for the complex entanglements that define our world (Pugh and Chandler, 2021). Within this, Jonathan Pugh and David Chandler point towards the emergence of 'patchwork' ontologies that are built upon the 'diversity, differentiation, interdependency, and feedback loops' of island life (Pugh and Chandler, 2021, p. 100). Emerging from 'the immanence of interactivity of life', these island-based ontologies directly oppose the characteristics of modern forms of resilience: individualism, technological progress, stability, equilibrium, and bouncing back.

Against the foundational thought of Plato and the economic essentialism of neoliberal subjectivity, contemporary forms of pedagogy are defined by an understanding of the self *with* the other. Suppose the student-centred, human capital subject is motivated by the desire to manage, order and prevent an encounter with the 'other'. In that case, the subject in education is defined by openness towards an encounter with the 'other' based on the assumption that the world is entirely unknowable, never static or controllable and never solely human (Pugh and Chandler, 2021). Within this, we argue for a shift in educational trajectories away from the resilience ontologies of modern education towards patchwork ontologies, where entanglements with the environment are to be understood generatively and creatively:

> Patchwork ontologies focus on patches or islands of creativity and refiguration forming in nodes or knots of assemblages across time and space. The ontological assumptions that all forms of being emerge through webs or networks of co-relation puts the emphasis on creative crossings and interconnections, meaning that new opportunities arise to see with and through these relations and co-dependencies.
>
> *(Pugh and Chandler, 2021, p. 78)*

Similar to Gert Biesta's call for a 'World-Centred education', patchwork pedagogies supplement modern modes of exchange in education that prioritise student-centered or curriculum-centred learning (Biesta, 2021). Arguing against the instrumentalisation of art education through economic and developmental logic, Biesta makes an existential case for art as a way for students to be in dialogue with *others and the world*, where the key educational question is 'whether what I desire is what I should desire, whether it is desirable for my own life, my life with others, on a planet that only has a limited capacity for fulfilling our desires' (Biesta, 2017, p. 70). Similarly, the Sophists had to 'take account of the diversity of opinions, the mediocrity of peoples' actual behaviour, and finitude in its lived reality' (Cassin, 2014, p. 117). Education, therefore, is necessarily political and aesthetic; it takes the form (*techne*) of teaching, but in that form, the 'here and now' shows itself (*kairos*) in the mixture of everyday life. Education is not about perfection or resilience but about opening our everyday lives to make their immediate necessity clear so we can act on that clarity. Such education does not try to solve the climate emergency by better fulfilling the desires of domination inscribed in the Platonic–Aristotelian line of educational thought. Instead, it lets us face the here and now of the everyday so we can be clear on what needs to be done in the reality of life as we live it with others and in an environment on which we depend.

Taking the patchwork structure as form, this publication supports difference and connection, stability and flux, dispersal and narration across ten chapters. They make a world together through practice, theory and resistance. In Chapter 1 Sharon Todd points towards the significance of events in our lives, from climate change to everyday events in education (black swans), that motivate subtle but essential shifts in perception. The great challenge posed by this chapter is how to curate the event in education for our students in such ways as not to instrumentalise it or to close it down as a curricular object but rather to construct an openness to the event through aesthetic experience. Key concepts in this argument are event as the rupture between old and new, post-event subjective experience, and event as an everyday tipping point. Importantly and consistent with this introduction, the event is presented as a disruption of the ideology and political reality of modernity/coloniality. The world that is ending is modernity and the centrality of human agency. Following this, we need to reimagine our being in the world *anew*, and one way to do that is through the aesthetic encounter.

Where Todd highlights the tension between modernist perspectives in education and the more entangled and relational perspectives emerging through contemporary Anthropocene discourse, Gert Biesta (Chapter 2) calls into question the idea of perspective or view as being at the centre of our new world problems. Through a deep historical and philosophical analysis of the emergence of the concept perspective, first in art historical discourse and then through philosophy, Biesta takes the Heideggerian 'characterisation of modern times as the 'age of the world view', as the central problem to be addressed through art and education in the Anthropocene. Oscillating around the tensions between the internal and external, the inside and the outside, Biesta arrives at a specific understanding of the education event as an external event, the event of 'givenness'. Developed by phenomenologist Jean-Luc Marion, givenness is similar in structure to interpellation, or the 'call', in that it prescribes a subject open to the newness of the event in the present, in anticipation of the event-to-come, 'that is to say, being able to interpret what is as what comes to us' (Marion, 2017, p. 39). For Biesta, the structure of 'givenness' breaks with the positionality and centrality of the human subject, that is, its 'modern worldview'. Decentred from this privileged position, from the binaries that elevated man above nature, the subject is, as suggested by Todd, now completely entangled *with* nature and, therefore, in a new world.

Extending these entangled enquiries, Carl Anders Säfström (Chapter 3) takes us on a journey back to antiquity and to the Greeks, where the roots of our contemporary pedagogical contract were carefully sewn (Too, 2000). For Säfström, through the Platonic–Aristotelian educational complex, we inherit our modernist privilege, at the centre of which are the binary exclusions that have enabled non-human subjects to be relegated to the recycling bin of history. One of the most violent expressions of this attitude in contemporary political theory came through Carl Schmitt's *The Nomos of the Earth*, where he rejected the ocean as utterly banal and inhuman compared to landmasses which support human progress, writing, 'The sea has no character, in the original sense of the word, which comes from the Greek *charassein*, meaning to engrave, to scratch, to imprint. … On the waves there is nothing but waves' (Schmitt, 2003, pp. 42–43; emphasis in original). Tackling these historical roots head-on, Säfström takes the plant as an excluded biological reality and an allegorical imaginary, spinning a web of associations between modernity and environment, consumerism and fascism, childhood story and Sami reflection. While these associations fold a performative aesthetic into a non-horizontal argument, the subversion of the foundational educational exclusion of Sophistic teaching by the Platonists is most pertinent for education during the anthropogenic climate crisis. Following this intervention, Säfström argues that it is only through the Sophistic practice of teaching as 'elsewhere' that we can attune our sensibilities to the otherness of the speaking plant. The 'elsewhere' of teaching as the event of teaching can move us beyond the social reproduction of human/non-human hierarchies in education to be in the world anew.

If Säfström aims to trouble the binary positions that have separated the human from the plant historically, artist and educator Barbara Knežević (Chapter 4) has transferred this dualist critique onto the material reality of art education through the concept of 'sculptural knowing'. In this analysis, the materiality of sculptural production is perceived as a tool to introduce students to empathic knowledge, knowledge that connects to deep time, ancestral practices and histories before scriptural language and rationalist knowing. This embodied pedagogy is positioned politically in opposition to the binary constructions of subject/object, form / inform and human/ non-human, but less discursively to the material realities of overconsumption and plastics pollution impacting the environment so significantly. For Knežević, sculptural knowing should be taught as an alternative epistemological framework for experiencing the world through objects, where direct contact and conversation with living, breathing materials enables an intersubjective being in the world, as opposed to our domination and control of the world. This deep and contemplative experience of sculptural knowing translates into a deep respect for the object as subject, its environment, and the cultivation of a pedagogical sensibility attuned to the patchiness of the climate system, unpredictable, dynamic and volatile. Beyond the neoliberal subjects' resilience to market instabilities that are also unpredictable, dynamic and volatile, Knežević's sculptural knowing proposes an ancient resilience connected to older systems of flux and adaptation for a sustainable life, what she calls 'clay memory'.

In 'Archipelago as Form: Education in a Post-climate World' (Chapter 5), Glenn Loughran reflects on educational form through the lens of archipelagic thinking. When considering the form of education, one should consider how the spatial dynamics of environment and context shape educational processes and subjectivity. Where the human capital subject of modern education remains the ideological blueprint for much educational theory and policy today, recent shifts towards Anthropocene thinking prompt a new conception of educational subjectivity entangled with the environment (Pugh and Chandler, 2021). At stake within this shift is an understanding of agency within education that is neither radically autonomous nor infinitely adaptable. Exploring the tensions between autonomy and adaptability, this chapter considers how central concepts developed within island studies can help think about these tensions in environmental agency. While agency in education often implies a shift in the teacher-student relationship, this chapter considers how the environment(s) act as agents within these educational forms. This analysis is supported by reflecting on the newly established archipelagic MA in art and environment in West Cork (2020). Emerging out of the BA in visual art on Sherkin Island (2004–2022), the MA in art and environment focuses on multiple geographic sites and educational themes.

The discipline of art history is also very well-placed to address the temporal shifts in environmental representation and to connect the past and the future in new ways that have become increasingly characterised by climate change discourses. As argued by Tim Stott in Chapter 6, the artefacts and art objects studied through art history are not passive receptacles of the past but vanish mediators of the past's relationship to the future via the present. As Alexander Nagel and Christopher Wood have suggested, they are 'a strange kind of event whose relation to time is plural' (Nagel and Wood 2010, p. 9). In this sense, the study of art objects as 'anachronic' and 'hauntological' events aligns the discipline of art history with the environmental humanities and its concern for entangled, multi-scalar temporalities. And yet, this slow shift towards an eco-critical art history has had to navigate inevitable tensions and nerve points around the urgency of the subject of study and the normativity of its educational and environmental contexts. Highlighting the conservatism at the heart of much curricular design in the field, Stott aims to address these tensions through a world-centred, eco-critical art history, one which uses Biesta's educational framework for a post-constructivist pedagogy to enliven and cultivate the urgency and integrity of the subject, inside and outside classroom settings. Contributing to the disciplinary formation of an eco-critical turn in art history, Stott also points to our institutions' internal habits and limitations to bypass the market ideologies that obscure these more challenging instabilities and their impact on the world to come.

Inspired by an earlier text by Biesta on art and education (Biesta, 2017), Juliette Clara Bortoldo (Chapter 7) draws out the significance of death, the dead and the past to provide pedagogical avenues for the present. Touching on the politics of mourning, Bortoldo shifts between the spectral, anachronistic capacity (Derrida) of the dead to teach towards the immanent and the material reality of teaching *with* the dead (Despret). This dialectic is navigated through the performance art project *Necropolis*, developed by Arkadi Zaides in Milan in 2021. Through her experience of the artwork, Bortoldo reflects on the ambitions of the project to capture the representation of the dead through multi-scalability. Shifting from the macro-technological lens of tagged cemeteries in

Google Maps to the entangled, granular experience of grave-side mourning *Necropolis* answers some of the questions posed by Bortoldos's text, specifically around the complexity of representation of the dead and the pedagogical implications of that representation. However, in the unpacking of those questions, Bortoldo introduces some new and important ones, specifically around the tensions and differences between art and education in their engagement with the dead, both as a historical trace, the impact of industrialisation on the present and a present, material fact, in the migrant flow of bodies resulting from the excess of global capitalism. The text contains several ethical questions that constitute a necessary ongoingness, an openness to the experience of the dead as living reminders of who we are and what we must do to live well in the Anthropocene.

The social and material realities of migrant life in Bertoldo's text take a more pragmatic and political turn in Paul O'Keeffe's 'Enabling Art to Become a Pedagogical and Therapeutic Tool in Refugee Camp Classrooms' (Chapter 8). Following years of frontline work in refugee education, specifically in the Kakuma refugee camp in Northwest Kenya, O'Keeffe reflects on the increasing need for educational investment and reform in such 'states of emergency'. As we continue to monitor, from a comfortable distance, the violent impact of climate change on the most vulnerable societies and countries, O'Keeffe brings us closer to the reality of such situations and, importantly, to the tensions and contradictions surrounding policy discourse in these 'states of emergency'. However, within this analysis, there are openings beyond the instrumental limitations of educational resources. O'Keeffe makes a strong case for the enhanced 'vitality' of educational experience in extreme contexts and precisely the psycho-pedagogical significance of art therapy to support a new generation of students to navigate the complexities of displacement, trauma and access in an increasingly violent and uncertain world.

How we prepare students to be in the world anew at a time of climate change is a complex, ethical challenge that artist and educator Fiona Woods takes on in Chapter 9, which reflects on a course that she runs at Limerick Technological University, titled 'Territories of the Self' (TOTS). Approaching the problem of art education in climate change through an intersectional lens, Woods brings together a spatial critique of neo-liberal capitalism, a historical reflection on aesthetic education and an attentiveness to embodiment in educational practice to argue for a performative pedagogy that can offer students the opportunity to 'examine the forms of social conditioning that they are willing to tolerate and accept'. Woods creates urgent educational spaces through the TOTS course for students to take ownership of the affective violence they experience daily through the mediatic onslaught of climate reporting. Importantly for this publication, the students in the TOTS course are perceived as producers rather than consumers of knowledge, specifically, producers and interpreters of the power of artistic research, which acts as a hermeneutic barrier and blockage to such systematic, symbolic violence.

Finally, through empirical studies of an English literature university course, a climate fiction course, the authors Leif Östman, Katrien Van Poeck, and Ellen Vandenplas (Chapter 10) make a strong case for developing theoretical frameworks to support the aesthetic event in education. The aesthetic event here is contextualised within the history of aesthetics as a necessary supplement to the instrumentalisation of contemporary education and to engage students in the 'difficult knowledge' of climate degradation. As defined by Deborah Britzman, 'difficult knowledge' must be navigated in the classroom rather than rejected, even though it may trigger students in different ways. Within this ethical dilemma, Britzman advocates for a practice of pyschopedagogy (Cho, 2009), where the teacher is trained and attuned to the already therapeutic character of teacher–student relations. However, for Östman, Van Poeck, and Vandenplas, the mediatic strategy to manage this process is not manifest in psychoanalytic technique but rather through art and its subversive capacity to sublimate the negative energy of reality in the climate crisis. Importantly for this publication, the authors switch between the power of art to cultivate creative agency for students in the classroom to teaching as an art form, equally complex and contingent, and dependent upon the teacher's creative and spontaneous response to the aesthetic dimensions of the event. In this sense, it is appropriate that the final chapter in the book aims to create a general equivalency between the power of art and aesthetics in education and the potentiality of the educative process as necessarily artistic and aesthetic. Where we have previously set out to stage the chapters in the book around a trans-critical dialogue between art and education and the different responses they illicit in the face of the event of climate change, we should also be open to reflecting on the commonalities between the disciplines, their characteristic vulnerability and evental capacity at a time of environmental and political disintegration.

References

Baptiste, I. (2001). Educating Lone Wolves: Pedagogical Implications of Human Capital Theory. *Adult Education Quarterly*, 51, 184–201. https://doi.org/10.1177/074171360105100302.

Biesta, G. J. J. (2017). *The Rediscovery of Teaching*. Routledge.

Biesta, G. J. J. (2017). *Letting Art Teach: Art Education after Joseph Beuys*. ArtEZ Press.

Bouchard, P. (1998). Training and Work: Some Myths about Human Capital Theory. In B. Spencer and S. Scott (eds), *A Canadian Reader in Adult Education*, pp. 33–68. Thompson Publishing Co.

Britzman, D. (2000). If the Story Cannot End: Deferred Action, Ambivalence, and Difficult Knowledge. In R. I. Simon, S. Rosenberg, and C. Eppert (eds), *Between Hope and Despair: Pedagogy and the Remembrance of Historical Trauma*, pp. 27–57. Rowman & Littlefield Publishers.

Cassin, B. (2014). *Sophistical Practice: Toward a Consistent Relativism*. New York: Fordham University Press.

Cho, K. D. (2009). *Psychopedagogy: Freud, Lacan, and the Psychoanalytic Theory of Education*. Palgrave Macmillan.

Consigny, S. (2001). *Gorgias: Sophist and Artist*. University of South Carolina Press.

David, C. (2022). Decolonising Resilience: Reading Glissant's Poetics of Relation in Central Eurasia. *Cambridge Review of International Affairs*, 35(2), 158–175.

Durkheim, É. (1956). *Education and Sociology*. Free Press.

Glissant, E. (1997). *Poetics of Relation*. University of Michigan Press.

Greenblatt, S. (2011). *The Swerve: How the World Became Modern*. W. W. Norton.

Ingold, T. (2024). Creation beyond Creativity: From the Novelty of Ends to the Renewal of Life. In H. Tsoukas*et al.* (eds), *Organizing in the Digital Age: A Process View*, pp. 19–35. Oxford University Press. https://doi.org/10.1093/oso/9780198899457.003.0002.

Jaeger, W. (1986 [1943]). *Paideia: The Ideals of Greek Culture, vol. III. The Conflict of Cultural Ideals in the Age of Plato*. Oxford University Press.

Jaeger, W. (1965 [1939]). *Paideia: The Ideals of Greek Culture, vol. I. Archaic Greece: The Minds of Athens*. Oxford University Press.

Jarratt, S. C. (1998). *Rereading the Sophists. Classical Rhetoric Refigured*. Southern Illinois University Press.

Kant, I. (2015). *On Education*, trans. A. Churton. Leopold Classical Library.

Lucretius. (1994). De rerum natura, trans. B. Inwood. In L. P. Gerson (ed.), *The Epicurus Reader*, p. 66. Hackett.

Marion, J.-L. (2017) *The Rigor of Things: Conversations with Dan Arbib*. Fordham University Press.

Nagel, A. and Wood, C. (2010). *Anachronic Renaissance*. Zone Books.

O'Sullivan, D. (2005) *Cultural Politics and Irish Education since the 1950s: Policy Paradigms and Power*. Institute of Public Administration (Ireland).

Paavolainen, T. (2017). Fabric Philosophy: The "Texture" of Theatricality and Performativity. *Performance Philosophy Journal*, 2(2), 172–188.

Pugh, J. and Chandler, D. (2021). *Anthropocene Islands: Entangled Worlds*. University of Westminster Press.

Reid, J. (2019). Securing the Imagination: The Politics of the Resilient Self. in J. Bohland, S. Davoudi, and J. L. Lawrence (eds), *The Resilience Machine*, pp. 29–42. Routledge. https://doi.org/10.4324/9781351211185.

Ryther, C. (2009). *The Possibility of Living Another Life: Reimagining Possibility in Radical Pedagogy Through Rancière*. PESGB Conference.

Säfström, C. A. (2020). *A Pedagogy of Equality in a Time of Unrest. Strategies for an Ambiguous Future*. Routledge.

Säfström, C. A. (2023). *Education for Everyday Life: Teaching as a Sophistical Practice*. Springer.

Schmitt, C. (2003). *The Nomos of the Earth*. Telos.

Tavory, I. and Wagner-Pacifici, R. (2022). Climate Change as an Event. *Poetics*, 93 (part A), 101600.

Too, Y. L. (2010). *The Pedagogical Contract: The Economies of Teaching and Learning in the Ancient World*. University of Michigan Press.
United Nations. (1945). United Nations Charter (Full Text). www.un.org/en/about-us/un-charter/full-text.
Vince, G. (2024). We Are in a Post-Climate Change World: Aaron Bastani Meets Gaia Vince. www.youtube.com/watch?v=ftEKVSZk7qU.

1

BLACK SWAN PEDAGOGY

Remnant Ecologies and Navigating the Difficulties of Disruption and Loss

Sharon Todd

Introduction

I drift slowly in the dark, attuned to the sounds of voices, both human and other, lights creating shifting forms I cannot quite recognise. It is eerie yet compelling, immersive yet strangely not of this world. The Botanic Gardens where I walk has become other to me, defamiliarised from the weekly pilgrimages I make there to admire the ever-shifting forms of plant life. Now it is night-time and as I tread upon the paths, sound, light and holographic images congeal into surprisingly new formations and I realise I keep stumbling into another element in this installation without even planning to. Ghostly and haunting, it is revelatory of what we are living through on this planet in uncertain times. In one installation I see an illuminated forest projected onto real trees, watching their growth from seedlings to giants and then their demise, as if in a dance of death. I am entranced by the movement and the play of light. But then the darkness comes—suddenly. The trees have vanished and the dark is heavier than heavy, absorbing as it does all life and light, and I become undone.

There is a strange combination of loss and beauty, disruption and playfulness at work here. As an educator I am in awe of Jony Easterby's *Remnant Ecologies* for the way it exposes, in a way conventional pedagogy cannot, the fragility of the world as we know it while also leaving us with some sense of its capacity to bear witness to what is no longer there—in its remembered—and indeed reimagined—beauty. One of the key issues facing educators in this time of environmental breakdown has precisely to do with navigating between disruption to and loss of life. With discourses of the Anthropocene suggesting a human-centred nihilism, on the one hand, and with commonplace appeals to hope for the future, on the other, educators are left with little resources for

DOI: 10.4324/9781003456636-2

thinking through alternative paths that not only avoid both extremes but that allow for being with the difficulties facing devastation on such a scale brings right here and now. For as environmental journalist Gaia Vince (2022) remarks, we are currently *living in* a post-climate change world; it is not some event coming to us in the near or distant future.

What this chapter proposes is a re-orientation of what education has to offer to these times of profound planetary change. To do this I turn to what art, in turn, can offer education to think through its limits and possibilities. I take as my starting point the ecofeminist artist Aviva Rahmani's (2022, p. 18) understanding that 'the best art is always a black swan, a surprising event', a term aptly borrowed from economist Nassim Taleb to indicate a disruptive event that changes fundamental perceptions of who we are. Akin to Jacques Rancière's (2006, p. 59) idea that art occasions 'a sensible or perceptual shock' that can create new communities of sense and existence, a black swan event offers us a way of moving beyond the poles of nihilism and hope that characterises current pedagogical thinking. Insofar as education is about staging encounters with elements of the environment (from books, screens, and maps to trees, water, and earth), which it shares with the practice of art (Todd 2023), how might we understand the nuances of these disruptive encounters in light of our current existential anxieties and environmental losses? That is, how do we educationally curate black swans that aspire to change, on the one hand, while attending to the feelings of loss that are part of experiencing ecological crisis in the present, on the other?

In response to these questions, I first outline in more detail the specifically educational problem in navigating through disruption and loss and draw out their connection to the matrix of modernity/coloniality (Machado de Oliveira, 2021). I then turn to explore the notion of 'black swan' and its link to art as a practice that breaks through conventional values and common sense informed by modernity/coloniality. In the third section, the paper explores what Easterby's *Remnant Ecologies* teaches us about our ambivalent connections with the more than human world and about pedagogical possibilities moving into the future. I then conclude with some thoughts on what I am calling black swan pedagogy.

Navigating Disruption and Loss: An Educational Issue

Disruption to life in these times of acute change is not only a personal, experiential issue but it plays a huge role in how to think of education itself. The aims of either socialisation or qualification seem quaint in the face of such colossal uncertainty about how we are to live on the planet. Couple this with loss—of species, of land, of ways of human life—and the enormity of what is happening on Earth hits us quite bluntly, leaving educators a little stunned and flailing about for firm footing. Often conceived as a coming to the 'end of the world', Anthropocenic discourse can turn to hyperbole to call attention to the extremity of the situation being witnessed. However, there is a cautionary tale in using such discursive constructs, becoming as they do almost fetishised as tropes of apocalyptic futures, since they risk not only the erasure of the differential ways human beings are affected in the present but also the threat to 'humanity' is largely seen through a Eurocentric lens and an increasingly 'technological management and control' mindset (Hine, 2023). By this I mean that there are already currently dire consequences for many humans and other species around the globe as a result of climate change with whom those of us in the global north are entangled. Such consequences neither lie in the future nor can they be resolved with recourse to 'western' science and technology without a change in how humans see ourselves as part of the environment we live in. As Lövbrand et al. (2015, p. 213) write, even 'the scientific Anthropocene narrative continues to portray nature as an object external to society with "natural" limits and tipping points that can be discerned, quantified and managed… As a consequence, humankind is both inserted into nature and re-elevated above it'. Thus the Anthropocenic call of finality reflects a continued logic of separation that underpins much of modernity—and all the attendant features of mastery it displays—not only over more than human life but over non-Eurocentric configurations of humanity itself.

As Mitchell and Chaudhury (2020, p. 310) query: 'it is often said the "end of the world' is approaching—but whose world, exactly, is expected to end?' Examining several scientific sources, they observe how BIPOC communities are continually positioned as 'threats to the survival of "humanity"'—and this is done through a number of (mis)perceptions: through the portioning out of blame for apparently high birth-rates; through the perception that their current deaths as a result of climate change are 'sacrificial' in potentially ushering in a new period of recovery; and through subsistence farming practices that are considered 'useless' in feeding the planet. What all this means for Mitchell and Chaudhury is that 'despite their claims to universality… "end of the world" discourses are more specifically concerned about protecting the future of whiteness' (Mitchell and Chaudhury, 2020, p. 310). As such, they charge, that some environmental solutions offered are not only re-enactments of colonisation of land and resources for BIPOC peoples around the globe (couched in the language of 'for their own good') but also reveal 'a profound anxiety about disruptions to—or reversals of—the linear concept of time that underpins European post-Enlightenment sciences' themselves (Mitchell and Chaudhury, 2020, p. 315). As such, educational responses to ecological disruptions and loss require a different approach to time and agency than the usual focus on shaping the present through one's hopes for the future, a point I return to below. This is particularly in order not to reinstall the colonial impulse whose tropes of progress, development and betterment have always been to maintain certain power relations that benefit a minority of humanity. Moreover the future-shaping narratives of 'hope' are also caught up within a logics of supremacy, which have marked modernity itself (Machado de

Oliveira, 2021; Chandler and Reid, 2019). What is meant here is that if 'hope' is considered to be a feeling or sentiment divorced from action, then it aligns with a Eurocentric sensibility that sees its lifestyle choices as not connected to how others are compelled to live their lives on this planet in servicing those choices. In this regard, hope is empty: it places attention onto an imaginary future which functions to keep us diverted from our interconnections—and the action this potentially leads to—in the present.

One way the current 'crisis' has been reframed is suggested by Vanessa Machado de Oliveira (2021) in seeing education as being engaged in the 'hospicing' of modernity. In her view, the colonial/modernity matrix is precisely the 'world' that is coming to an end. Rather than embrace Anthropocenic doomsday scenarios that see 'the world' (or nature) absolutely (and singularly), Machado de Oliveira emphasises that it is 'the world as we know it', not 'the world' per se, that is at stake. What we are witnessing is the long, slow and painful death of modernity itself whose extractive and separationist logics have become self-consuming. As Hine (2023, p. 10) also cautions: 'unless we can inhabit that distinction [between the world as we know it and the world itself], we will end up defending the world as we have known it at all costs, no matter how monstrous those costs turn out to be'. Thus, the claims to the 'end of the world' can lead to defensive positions to preserve lifestyles (both modern and colonial) that actually are at the root of the climate emergency in the first place.

For Machado de Oliveira, to fully face the end of the world as we know it requires us is to see and experience how we are all encompassed within the modernity/coloniality frame to lesser or larger degrees. And it requires, as she emphasises, a specifically *educational* process in order to facilitating this facing. Part of the educational work that needs to be done is to reconnect the existential elements of ourselves that we take to be second nature with the state of the planet and its inhabitants. 'Decluttering' beliefs and values that serve a master who has never been viable (e.g. desires of consumption, supremacy, and efficiency projected onto the natural world as well as colonised others) involves seeing ourselves as part of the loss of modernity, existentially speaking. She refers to this process as 'depth education'—an education that seeks to uncover the ways each of us as a subject is both formed by and can transform in turn the stories, affects, values, and logics of modernity. Depth education, she writes, 'is an orientation toward activating capacities and dispositions that can enable us to hold space for difficult and painful things, and to sense, relate, and imagine otherwise as we face the end of modernity or the world as we know it' (Machado de Oliveira, 2021, p. 43). Hence Machado de Oliveira sees that sensory and imaginary processes (beyond cognition and intellectual work) are necessary to creating an educational response up to the task of facing environmental breakdown and all the social, political, ethical and existential realities that it brings into relief.

It is important to be clear here that educational responses to disruption and loss are not about psychologising or individualising the problem of ecosystem collapse, neither are they simply about 'delivering' more information for students to digest, with the assumption that this will precipitate a rational form of action. Rather, it is about reorienting what education does and can do in a way that bridges the existential, sensory subject with the lived realities it is facing and the complexity of the affect that accompanies such bridging. What is key here is that depth education is not simply that is carried out by an individual, but is a collective process focused on the intersection of where the singular subject meets the larger world of which they are a part. To take seriously the idea that the senses and affect matter in reimagining the world otherwise means taking an aesthetic and public approach to our educational endeavours instead of the usual cognitive one.

In his paper 'Beauty in the Darkness', Ramsey Affifi (2020) focuses on the importance of beauty in environmental education. Not in terms of the romantic notions of the sublime or that which strikes awe in the beholder, but in terms of the vulnerability of what we behold, what Affifi refers to as the 'beauty of the vulnerable'. Vulnerability can take the form of the perishing of a bee, the death of a flower, the languishing of a beloved pet. This is a fragile state that loss (or potential loss) ushers in, heightening our aesthetic sense and calling forth another kind of attention—and action. The beauty of the vulnerable also allows us to navigate the difficult terrain of loss without the simplistic assurance of hope. For Affifi, there is a particular tension at stake for the environmental educator since, on the one hand, focusing on the devastation and horror of extinctions can lead to forms of despair, anxiety, and apathy, and on the other, glossing over the grim details and suggesting that the problem will be solved also runs the risk of apathy and lack of attentive action. Affifi suggests taking an alternative path: 'Opening to and participating in beauty skirts this dilemma because beauty is most compellingly intuited when we see evermore clearly the situation we are faced with' (Affifi, 2020, p. 1132). Importantly, the motivation for action comes neither in despair nor hope, but in an attention to the fragile state of being. Beauty here signals a form of what I think of as a form of affective attention—a tending that is felt and experienced, moving us to act not despite of fragility but because of it. In this, there is sense that when we recognise the disruption and loss in life, something else can happen, something can be introduced that acts not like a soothing balm (such as we find with hope) but as a point of compassion and care that is at once poignant and fragile. It is from this encounter with elements in our environment, be these trees, flowers, birds, soil, insects, water or air, that the sensory relation becomes a relation with beauty. At the point where disruption appears, there opens a space for beauty to enter.

Beauty for Affifi is largely discussed in terms of the environment, and not art (which is often its conventional cultural home). But in response to his question, 'how can we work with children [and adults] to help them invest into these vulnerable vulnerabilities?' (Affifi, 2020, p. 1135) it seems to me that art is uniquely positioned to create such possibilities. This is particularly so if we consider education to be about a constant curation of encounters that are both sensory and educational (Todd, 2023). That is, seeing education as an aesthetic process allows an exposition of the vulnerable that generates affective and sensory responses. On this view, there is something in the act of education that lends itself well to facing disruption and loss and to undoing modernity's death grip through artistic experiences. Moreover, the question of 'whose world is expected to end' can be kept at the forefront of aesthetic encounters since the way they operate at the level of vulnerability can open us up to investigate how we are caught up in regimes of supremacy—regimes that articulate whose world counts—not only among humans but also between human and more than human others. That is, they can begin the work of unravelling our commonplace assumptions about ourselves and our place in the world. There is, I argue, something to be said for the staging of aesthetic encounters through art that can also pierce through the carapace of business as usual and begin to expose us to what we feel right now, in the present, and that allow us to unfold toward a future that reimagines the world differently.

'Black Swan' Events

The economist Nassim Taleb (2007) draws our attention to a story of hubris—the sighting of a black swan by British colonists in Australia came as a shock since they were falsely convinced that all swans were white. It was the power of false universalisation of the category 'swan' and which prevented them from imagining a swan of a different hue. For Taleb this surprising event for the colonists represents the 'fragility of our knowledge' (Taleb, 2007, p. 10) and the ways in which certainty limits our ability to predict the world. It is perhaps a telling part of the story that the impulse to universalise went hand in hand with the impulse to colonise—that is, the ways we perceive can be bound up with powerful myths and imperialist modes of being that we simply take for granted. Extrapolating from this example, Taleb identifies three characteristics of what he coins a 'black swan event' (i.e. an event that carries tremendous social, economic and political upheaval):

1. Rarity—that it must be an outlier since the past has not predicted its coming into being.
2. Extreme impact—that it brings profound impact causing disruption, loss and reorganisation.
3. Retrospective predictability—that we can explain its effects only after the fact.

(Taleb, 2007, p. 10)

For Taleb, this makes '*what you don't know* far more relevant than what you do know' (Taleb, 2007, p. 11). Despite the conservative politics that Taleb embraces (which do not take into account the devastating effects modernity/coloniality has upon the planet, including his failure to challenge the story of colonisation that informs his concept), there is something about the element of unpredictable shock that is central, in my view, to understanding the connection between art and education and the way they can together navigate through disruption and loss—without losing sight of the dynamics of supremacy that have brought about ecological collapse in the first place.

Drawing on Taleb's formulation of the 'black swan' in her memoir *Divining Chaos: An Autobiography of an Idea,* ecofeminist artist Aviva Rahmani explores what drives her activist-oriented work. Rahmani is interested in disruptive events 'that change fundamental perceptions' (Rahmani, 2022, p. 17). Rather than see such events on the same scale as Taleb does, Rahmani sees that relatively 'smaller' events have system-wide repercussions. And it is in this light that she claims that the 'best art is always a black swan, a surprising event' (Rahmani, 2022, p. 17) one that alters the rules of any given system. A black swan in this context is indeed rare and impactful (even if its effects might not be retrospectively predictable) and, for Rahmani, has less to do with the artwork itself than with the gesture with which it is made; that is, ecofeminist art carries with it a movement to alter the environment as well as viewer/participants' perceptions of it. It does not need to be on par with the pandemic or economic crash of 2008 to make a difference, to become a fully-fledged 'Black Swan' in Taleb's sense. Instead, Rahmani is resolutely committed to seeing art as having an effect that both is personal and also initiates change, flowing out from the point of its insertion into the world to transform both the physical and relational landscape in which it emerges.

Drawing more specifically on chaos theory and developing her own understanding of what she calls 'trigger point theory', Rahmani outlines the ways in which art, aesthetic language and intuition can precipitate a 'tipping point' that shapes a system of interactive relationships. Thus art can insert itself into a system and thereby shift the terms/rules/regulations that govern that very system. She writes, 'there will always be a small point of entry into any chaotic system' (Rahmani, 2022, p. 26) akin to an acupuncturist's needle in the system of energy that flows throughout the human body. To declare that even a relatively small trigger point can become a black swan, means a work of art is neither unimportant nor incidental to the environment out of which it is created—to see it otherwise would be to ignore the web of interactive relationships that constitute our coexistence with human and nonhuman others. Artworks that engage in practices that intervene in the environment (for example those that work with gardening, tree planting, food production and rewilding are more obvious ones) are indeed disruptive events that change the physical landscape, how we perceive it, and our own relations within it. Rahmani takes this a step further:

> trigger point theory is [itself] an artwork, a drop of hope flying in the face of despair, in defiance against apocalyptic probabilities, dysfunction, and division, a complex adaptive model for the ecological crises ahead, a vision of what art could contribute to change, the ephemeral sound of one hand clapping.
>
> *(Rahmani, 2022, p. 286)*

Rather than provide a 'drop of hope', however, I wonder if trigger points do more than this in suggesting that the disruption they cause actually enables us to stay both with our implication in events as they unfold, and with our vulnerability to those events at the same time. That is, an artwork provides an encounter that opens up a space that is unsettling because it invites us to perceive our world otherwise and enacts a loss of certainty in so doing. As Rancière (2006, p. 59) phrases it, art can create a 'perceptual shock' that causes us to shift our 'common' sense—interrupting into hegemonic modes of perceiving and inducing new modes of sensibility. Disruption and loss are therefore crucial to transformation. As Affifi emphasises in relation to beauty, it is the unexpectedly fragile circumstances of our situation that opens up the possibility for action.

Remnant Ecologies

I return at this point to Easterby's *Remnant Ecologies* in order to see what we can learn from the way it potentially acts as a trigger point into continued systems of environmental degradation and discuss the ways this is pedagogically significant. Composed of light, video, sculpture, and soundscape, the 13 installations that make up *Remnant Ecologies* were positioned throughout the National Botanic Gardens in Dublin in 2022 (with a return of the work in 2023). Audiences meandered in the dark through pathways linking each installation to another, with the echoes of the previous installation still lurking as one approached the next one. Some were monumental in scale, such as *Vegetative Succession*, which was the piece I wrote about earlier in the introduction—a roughly 50-metre-long video projection of the life and death of a forest. Others were more intimate, relatively speaking given the vastness of the gardens, such as one comprised of the mechanical movements one finds in music boxes which were encapsulated in vitrines made to look like bird cages; the vitrines were lined up on either side of a pathway and emitted, at various tempos and tone, the call of cuckoos whose song can no longer be heard in nature near the artist's home in Wales (see Figure 1.1).

As its title suggests, the immersive, large-scale installation is indeed about remnants—fragments and pieces that comprise what is 'left' in this time of environmental breakdown. What remains when one can no longer hear the actual song, see the actual trees, touch the actual seeds? And yet, at the same time, this is precisely what the artwork enables us to do by proxy, presenting

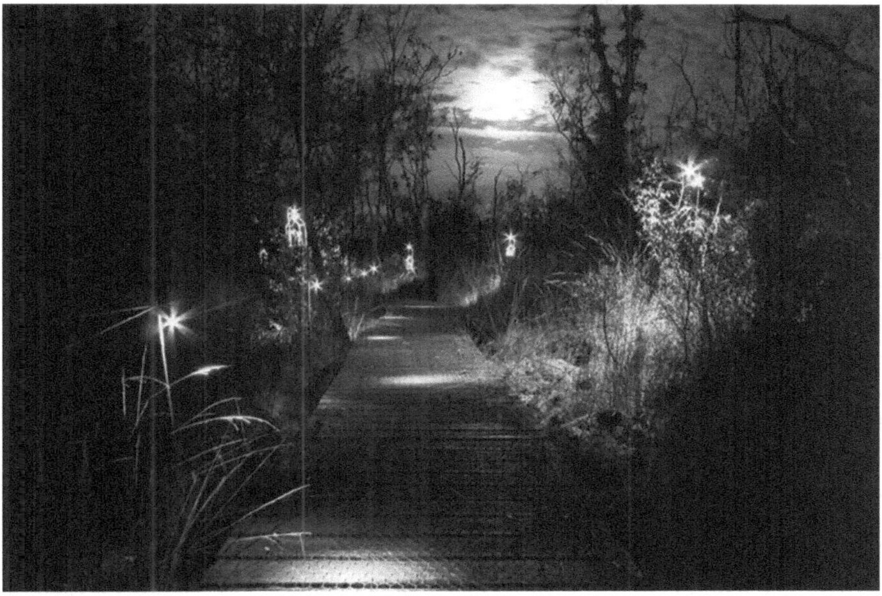

FIGURE 1.1 Jony Easterby, *Cuckoo 1.*
Source: retrieved 2 March 2024 at www.jonyeasterby.co.uk/portfolios/remnant-ecologies-2/

us with vestiges of plant and animal life that we nonetheless suddenly sense as loss and presence simultaneously. The installations invoke presence through the profound encounter between bodies and video/sound. The visual and auditory language inserts itself into the 'real' time of the gardens to disrupt our temporal ordering—how I am hearing something that is gone? How does a video dance of an emerging and dying forest conjure a sense of time condensed? What these remnants do is not simply re-present something that is no longer; rather they 'presence' loss and disruption in the very moment of encounter. We become part of the ecology of these remnants. Neither, therefore, a 'lesson about the future' nor a simple lamentation for the past, the artwork reminds us that the felt sense of loss is *now,* and the repercussions of what is dying and disappearing are vibrating outwards from those points of contact. At the same time that we are witness to the demise of those birds and trees, there is a strange and uncanny sense of emergence with them; that we are present together in strange slip of time. To what extent do- or can—educational encounters enter a similar temporal landscape?

Black Swan Pedagogy

What *Remnant Ecologies* brings to the fore is the way curating encounters with disruption and loss is itself a disruption into the everyday fabric of life which compels us to see our relations anew—as part of a relationscape (Manning, 2012) through which we emerge together to form new constellations of becoming. Returning to how to think about education in these times not only in the context of anxiety for the future, but also in the context of the actual devastation currently taking place, I highlight here the importance of advocating for a pedagogy that takes seriously the curation of black swan events. Not in Taleb's sense of having meteor-like world impact, but in Rahmani's sense of ushering in change that ripples outward to effect a shift in the current flow of common sense. Black swan pedagogy would act through the idea of creating trigger points—not for their own sake to produce simply a 'shock effect' (for example, as one might find in the rifling off of long lines of scary statistics about tipping points and end of the world scenarios)—but in order to be able to sit with spaciousness in confronting our own implication in the environments we live *in* (and not only the ones we live 'from', as Bruno Latour, 2017, puts it). A black swan pedagogy is about enabling interconnection between human and more than human others, seeing the existential (i.e., human loss and anxiety) as bound to the more than human world. In this, black swan pedagogy is a form of depth education, whereby one's sensibilities are untangled through a painful process of ridding ourselves of those modern and colonial illusions of separateness and exceptionalism that keep the system of environmental harm alive. And it is a formation of possibility in allowing us to move beyond our own limitations of mindset to generate new ideas and forge new connections.

In staging trigger points, educators, like artists, do not re-present the world for others, but presence the world in ways that seek neither to overwhelm nor to 'sugar-coat' (to quote Affifi here) environmental breakdown; instead trigger points allow for a range of affect, sensation and perception to emerge, bringing to the fore the multiple and ambivalent ways one singularly experiences aspects of what is currently happening to the environments we are enmeshed in/with. I think of Affifi's (2020) examples here, of having students tend gardens or care for animals not in order to 'instruct' them 'about' plants or animals, or even to cultivate a 'love of nature' but to enable them to experience the weight of life (and sometimes death) so that beauty can emerge in the fault lines of living and dying. But as I mentioned earlier, it is not just about a direct relationship with nature that is relevant for education. Instead it is in the ways we, as educators, *curate* encounters with and for students so that they resonate with the aesthetic, sensory dimension of their experience as entangled subjects of the world. For new worlds cannot be taught as if they are like old ones, they must be made out of new sensibilities, new perceptions and ways of experiencing which, as Rahmani suggests, is what art does so well—and what education can do as well albeit under different circumstances. In this sense, black swan pedagogy is itself a form of being artful about navigating our vulnerabilities in this time of crisis in order to generate alternative worlding practices than the ones we have inherited. After all, it is what we don't know—and not the certainties of the past—that is most relevant for living in a post-climate change era.

References

Affifi, R. (2020) Beauty in the darkness: aesthetic education in the ecological crisis, *Journal of Philosophy of Education*, 54 (4), pp. 1126–1138.

Chandler, D. and Reid, J. (2019) *Becoming indigenous: governing imaginaries in the anthropocene.* London: Rowman and Littlefield.

Easterby, J. (2022) Remnant Ecologies. www.jonyeasterby.co.uk/portfolios/remnant-ecologies-2/, accessed March 2, 2024.

Hine, D. (2023) *At work in the ruins: finding our place in the time of science, climate change, pandemics and all the other emergencies.* Chelsea, VT: Chelsea Green Publishers.

Latour, B. (2017) *Facing gaia: eight lectures on the new climatic regime.* London: Polity Press.

Lövbrand, E., Beck, S., Chilvers, J., Forsyth, T., and Hedrén, J. (2015) Who speaks for the future of Earth? How critical social science can extend the conversation on the Anthropocene, *Global Environmental Change* 32, pp. 211–218.

Machado de Oliveira, V. (2021) *Hospicing modernity: facing humanity's wrongs and the implications for social activism.* Berkeley, CA: North Atlantic Books.

Manning, E. (2012) *Relationscapes: movement, art, philosophy.* Cambridge, MA: MIT Press.

Mitchell, A. and Chaudhury, A. (2020) Worlding beyond 'the' 'end' of 'the world': White apocalyptic vision and BIPOC futurisms, *International Relations*, 34 (3), pp. 309–332.

Rahmani, A. (2022) *Divining chaos: the autobiography of an idea.* New York: New Village Press.

Rancière, J. (2006) *The politics of aesthetics: the distribution of the sensible.* London: Continuum.

Taleb, N. N. (2007) *The black swan: the impact of the highly improbable.* New York: Random House.

Todd, S. (2023) *The touch of the present: educational encounters, aesthetics and the politics of the senses.* Albany: State University of New York Press.

Vince, G. (2022) *Nomad century: how climate migration will reshape our world.* New York: Flatiron Books.

2

AS IF THE WORLD IS WAITING FOR OUR OPINION

In Search of a (Re)configuration

Gert Biesta

Introduction

In this chapter I am after something that may be quite difficult to achieve, if it can be achieved at all. Some may want to characterise my ambition as paradoxical, or they may even want to argue that the difference I seek to make is not as much of a difference as I may consider it to be. I am very willing to engage with such considerations, but would like to ask the reader for some patience and for an opportunity to set out my 'case', so that it becomes possible to judge how far I have managed to go. Such a judgement is, of course, for the reader to make; the only thing I can do, is to provide as much detail and clarity as possible.

The difficulty of the ambition I have, is that I am not proposing a different view or perspective, and also am not asking readers to consider or adopt a different view or perspective. Rather, there is an invitation in the paper to 'locate' oneself differently—a change of heart perhaps more than a change of mind—and perhaps also an invitation to 'conduct' oneself differently. In the title I call it, clumsily, a search for a (re)configuration. As will hopefully become clear in what is to follow, I am not arguing for a new or different way of viewing the world, and even less so for a new worldview, but am rather asking what it would require from the self to meet the world differently and, through this, also meet oneself differently. If it helps: I am not trying to articulate a different 'third person perspective'—a view from the 'outside', so to speak—but am trying to consider the 'first person singular', to use a phrase from Alfonso Lingis (2007).

The guiding idea of this chapter is that, as a response to the contemporary environmental crisis, the search for new perspectives, worldviews and ontologies may be problematic. The reason for this lies in the intuition that the very

DOI: 10.4324/9781003456636-3

idea that we can have 'a perspective on' or 'a view about' the world may itself be part of the problem that needs to be overcome. This is first and foremost because it implies a configuration in which the world—natural and social—is 'approached' as an object for our views, perspectives and opinions and, in this way, literally becomes objectified. It is also, because in this configuration those who have views, perspectives and opinions stand, in a sense, outside and before the world, which at the very least raises the question *where* they actually stand and what would make it possible for them to stand there. Rather than looking for new perspectives, it is this configuration *itself* that needs to be overcome. In this chapter I explore how this might be done and what this might mean for education in a time of crises.

I will develop my line of thinking in the following steps. I start with a discussion of the idea of 'having a perspective' and will show, taking inspiration from a remarkable essay by Martin Heidegger, that the idea that of 'having a perspective' is not a neutral, trans-historic category, but a configuration with a very specific origin and 'location'. Heidegger refers to this as 'the age of the world view' (or, in another translation: 'the age of the world picture'). I identify a number of key problems with the idea of what I will refer to as 'perspectivism' and suggest that to address these problems we shouldn't be looking for new and better perspectives, but rather for what, for the moment, I will refer to as a different 'configuration'.

In a second step, I will outline what such a different configuration may look like, bearing in mind that even a phrase such as what something may 'look like' is part and parcel of the 'logic' of perspectivism. For this I take inspiration from phenomenology. This is, of course, not the only 'route' that can be taken, but I find it a meaningful and helpful one for what I seek to explore. In the third section of the chapter I discuss ideas from Husserl, Heidegger, and Levinas. This is not only to show how they try to bring the world 'back in', but also in order to explore how the self 'returns' in relation to this. Husserl and Heidegger are both helpful where it concerns the world, whereas Levinas is crucial for highlighting the way in which the self can show up in this altogether different configuration.

In the fourth section of the chapter I turn to the work of Jean-Luc Marion, who also firmly stands within the phenomenological tradition. Marion has devoted a significant part of his scholarly career to an investigation of the phenomenon of 'givenness', particularly in order to 'grasp', if that is the right word—it actually isn't—the givenness of the phenomenon, that is, the way in which the world first of all gives itself (to us). This, as I will try to explain, also allows for the self to 'appear' differently than as 'world-viewer' or 'perspective-haver'. Key in all this is how the world appeals to us, irrespective, so we might say, of our own views, opinions, and perspectives.

In the concluding section I draw the main lines of the chapter together and suggest briefly what this might entail for an education that seeks to break away from the 'pull' of perspectivism. Because I try to do quite a lot of (philosophical) 'groundwork' in this chapter, the discussion of education will be short and concise, but will hopefully provide those readers who find the trajectory of this chapter meaningful with a number of 'pointers' for further development.[1]

Taking a Perspective: The Age of the World View

The idea of 'taking a perspective' has become a very common and everyday 'gesture'. In conversations people often preface what they say with a 'from my perspective', and also frequently ask others about their perspective. Scholarly work often identifies itself in terms of a particular perspective—a European perspective, a Nordic perspective, a perspective from the global South, a post-structural perspective, a feminist perspective, a posthuman perspective, and so on. There is something good about all this, as prefacing what one has to say with a 'from my perspective' or 'from this perspective' clearly puts the cards on the table, and takes away any suggestion that what is being said is 'the view from nowhere', that is, the final truth or the total(ising) picture.

The language of 'perspective' has become so common, that it is easy to forget that the very idea of 'perspective' has a specific history and origin, and that it is, in that regard, a little less innocent or universal as one might assume. The modesty implied in the 'from my perspective' or 'from this perspective'—which is of course good and helpful—runs the risk of hiding what is going one when the idea of 'perspective' is invoked.

One history of what, for now, I will refer to as the idea of 'perspectivism', goes back to fifteenth-century Florence and to Filippo di Ser Brunellesco (1377–1446), who is generally seen as the inventor of linear perspectival drawing (see, for example, Edgerton, 2009). Brunellesco was the first who managed to represent three-dimensional structures with a 'real life' perspective onto a two-dimensional surface. While Brunellesco's invention can be seen as a major *technical* achievement in light of a long history of non-perspectival visual art, it is not just or not only a matter of technical mastery but also, and perhaps first and foremost, a *reconfiguration* of the (view on the) relationship between individual and world.

What is first of all distinctive about perspectival drawing is that it defines *a precise standpoint* for the observer. It defines the point where the observer stands as the point from where the observer is to observe what is to be seen, and to observe it in the 'right' perspective, so to say. This also means that perspectival drawing defines *a field* in which the observer is positioned as a subject-observing-an-object-from-a-particular-standpoint. This is very different from, say, Egyptian art or religious icons, where there is no defined point for the observer and, therefore, a very different field of relationships. In non-perspectival art the observer can be anywhere, which also means that the observer is nowhere, or to be more precise, the observer is *nowhere in particular*.

Viewed in this way, the invention or 'arrival' of perspectival drawing thus marks a wider reconfiguration of the relationship between human beings and the world (on this see, for example, Gruwez, 2009). It is a reconfiguration in which human beings appear as observers-of-the-world from a particular standpoint, and where the world appears as that-which-is-being-observed by the observer from their standpoint. Individuals thus gain the identity of observers or 'world viewers' (see below for this term), and encounter the world—natural and social—as object to observe from their particular standpoint or perspective.

In one sense, therefore, perspectivism is a gain, because it allows for a defined position for each individual vis-à-vis the world. It thus brings about a kind of *freedom* for each individual. We can characterise this as the freedom of observation, the freedom of a standpoint, and also as the freedom of (having a) perspective. But perspectivism also entails a loss, because in it the individual arrives as spectator-of-objects, the individual arrives as 'world viewer', which not just turns the world into an object that individuals have perspectives *on*, but also puts the individual in a particular relationship with the world or, more precisely, puts the individual at a distance from the world. The gain of the individual's freedom, thus also comes with losses—first of all a loss of the world, which is turned into an object, but also a loss of location, so to speak, as it is not immediately clear where the individual actually stands.

One question this raises is whether perspectivism should be characterised as a typical *modern* view of the world or, in slightly more technical language, as a typical modern world view or *Weltanschauung*. In a fascinating essay from 1938 called, in one translation, 'The Age of the World Picture' (Heidegger, 2002) and, in another translation, 'The Age of the World View' (Heidegger, 1976)—in German: 'Die Zeit des Weltbildes'—Martin Heidegger has suggested that what is modern here is *not* this particular *view* of the world, but rather the very idea of 'the world understood as view' (Heidegger, 1976, p. 350) or 'grasped as picture' (Heidegger, 2002, p. 67). It is not, therefore, that there is a shift from, say, a medieval world view to a modern world view. Heidegger rather suggests that it is 'that the world as such becomes a view' which is 'the distinguishing mark of modern times' (Heidegger, 1976, p. 351). Hence his characterisation of modern times as the 'age of the world view'.

Heidegger refers to what is happening here as a 'transformation' (see Heidegger, 1976, p. 350) in a similar vein to what, above, I have described as a reconfiguration. This is so, because it is not that an already existing individual changes their view about the world, but rather that individual and world emerge *together* in a new relationship and with new identities. As Heidegger puts it: 'That the world becomes a view is one and the same process with that by which man … becomes *a subjectum*' (Heidegger, 1976, p. 352; italics in original). And it is only now, 'for the first time', so Heidegger writes, that 'there is such a thing as a viewpoint of man' (Heidegger, 1976, p. 352).[2]

Heidegger also argues that it is in the 'interweaving of both processes—that the world becomes a view and man a *subjectum*' that 'the basic process of modern history' becomes visible (Heidegger, 1976, p. 352). He describes this process as a kind of double move in which 'the more objective the object seems to be, the more subjectively … does the *subjectum* rise up' (Heidegger, 1976, p. 352). Because the subject becomes more and more 'subjective' the more the world becomes (a) view, this development ends up in what Heidegger characterises as a 'battle of world views' (Heidegger, 1976, p. 354). In this battle, Heidegger writes, 'man fights for the position in which he can be that existent which sets the standard for all existence and forms the directive for it' (Heidegger, 1976, p. 353). The standpoint is therefore not just an epistemological category, but has political significance as well.

The first thing to reiterate is that if it is only in the 'age of the world view' that the world becomes something to view and the individual becomes the one who views the world from their unique, individual and, in a sense, free standpoint, then we can no longer see the idea of 'world view' as a neutral, a- or trans-historical category. It makes no sense, then, to speak about, say, a Greek, a Roman, a medieval, an Eastern, or a Western world view (and also not, as mentioned, a modern world view). We should rather conceive of these as *different configurations* or constellations—or if one wishes: human-world assemblages—of which 'world view' is only one possible option.

It also no longer makes sense, then, to speak about a Christian or Islamic world view or about religious world views more generally, because doing so would pull such religions into a very particular modern and, so I wish to suggest, secular configuration.[3] The same holds for the idea of indigenous world views as, again, to characterise 'indigeneity' as a world view would immediately pull it into the (modern, Western) 'logic' of perspectivism. This would most likely be a denial of the very 'claim' one seeks to articulate with that notion, namely that 'indigeneity' has its own 'right' to exist and is precisely not dependent upon a 'logic' of 'framing' from 'elsewhere'.

The second thing to reiterate, is that when 'the world becomes picture' (Heidegger, 2002, p. 67), the world *itself* is, in a sense, lost. This also means that the individual in a sense 'disappears' from the world, because as 'world viewer' the individual now stands *outside and before* the world-as-object-of-observation, as object-to-have-a-perspective-on. This shift is exactly the trajectory of modern philosophy, in which the question how the human being can still be 'in touch' with the world and gain reliable knowledge of it, not just became the main *question* of modern philosophy in the guise of epistemology, but also became the main *problem* for modern philosophy. The latter is epitomised in Kant's notion of the 'Ding-an-sich', which is the phrase he introduced to refer to reality which, in itself ('an sich'), is entirely unknowable and, more importantly, unreachable.

The problems of the age of the world view are, however, not merely philosophical and not merely of the past. On the contrary, we are currently in the middle of a number of significant problems stemming from the logic of perspectival reason. One is the idea of 'post-truth', which clearly stems from the suggestion that knowledge and truth are a matter of individual perspectives 'on' the world, with no outside arbiter. Similarly, identity politics stems from the idea of identity as the perspective one has on oneself (as individual or group), plus the suggestion that everyone should have the right to one's own perspective on oneself. While this can easily result in a 'battle of world views' at individual and group level, such a battle is currently also waging at a global level, with devastating consequences.

In the domain of values, the logic of perspectivism has resulted in the conflation of values with (individual or collective) preferences, which is a very different way to think about value and values than from the acknowledgement that something or someone may *have* value, be of value or be valuable in itself. One could also argue that the recent proliferation of ontologies and '-isms' such as posthumanism, new materialism, or relational ontology, is an instance of the logic of perspectivism, stemming from the assumption that also at a meta-level we are just free to articulate our own perspectives on the world, that we are just free to tell the world what it is or what it ought to be.

The logic of perspectivism also speaks to the current ecological crisis, because if the world is just an object for us to view and have a perspective on, it becomes quite easy to 'forget' about the world itself and about its integrity, and just do with the world what we want to do with it, without considering the limits and limitations the world may want to set. A more 'extreme' version of the objectification of the natural world is the denial of reality altogether, which, more than just claiming the right to one's own perspective, is actually to say that the world itself doesn't really matter any longer.[4]

Phenomenology: A Different Perspective or Altogether Different?

If perspectivism entails a certain 'loss' of the world, particularly in that it turns the world into an object, and also entails a certain 'loss' of the self because it positions the self outside and before the world, then there is the question whether it is possible to 'return' world and self in some way. I hope that it will be clear from what I have said so far, that this is precisely *not* the search for a new perspective, but perhaps better captured as the search for a new 'configuration'. In this section I will discuss, albeit briefly, some key insights from thinkers who, in different ways, belong to the tradition of phenomenology. I do so, because I wish to show that what unites these thinkers is their ambition to configure the field in which world and self can appear radically different from how this is done in the age of the world view, that is, according to the logic of perspectivism. Phenomenology thus provides a way in which world and self can return and be returned.

This ambition is clearly articulated in the 'motto' for which the word of Edmund Husserl is often known, which is the suggestion that we need to go 'back to the 'things themselves''(Husserl, 2001, p. 168). The Greek word '*phainómenon*' (φαινόμενον) does indeed mean 'thing appearing to view' and already in this phrase we can see that phenomenology wishes to focus on things—in the widest sense of the word—that appear to view (or more precisely in the case of Husserl: to intuition), rather than on the mental experience of things. Phenomenology thus puts the thing in the centre and in this way 'performs' the opposite gesture from perspectivism, where it is always the viewer who is in the centre.

Whereas Husserl's ambition was to go back to the 'things' themselves—and here I am only referring to Husserl's ambition, not to the question whether he managed to achieve this ambition (for a critical discussion see Carman, 2006)—Martin Heidegger proposed a rather different 'agenda' for phenomenology, namely an investigation into how 'things' exist or, in his own words, an investigation into the *being* of beings (in German: 'das Sein des Seienden', often written in English as the Being [with a capital B] of beings).

As Heidegger put it:

> *For Husserl*, phenomenological reduction ... is the method of leading phenomenological vision from the natural attitude of the human being whose life is involved in the world of things and persons back to the transcendental life of consciousness. ... *For us* phenomenological reduction means leading phenomenological vision back from the apprehension of a being ... to the understanding of the being of this being.
>
> *(Heidegger, 1982, p. 21)*

It is this 'ontological difference', that is, 'the differentiation between Being and beings' (Heidegger, 1982, p. 17) that is at the heart of Heidegger's phenomenological 'project'. Philosophy, for Heidegger, thus becomes 'the theoretical conceptual interpretation of being, of being's structure and its possibilities' (Heidegger, 1982, p. 11).

If in the logic of perspectivism the individual is in the centre of things, actively projecting a perspective onto the world, the way in which the human being shows up in Heidegger's work is, again, entirely different and, in a sense, the total opposite of this configuration. Heidegger emphasises qualities such as passivity and receptivity, which is also captured in his suggestion to refer to the human being as 'Dasein', which can be translated into English as (a) 'being-there'. The Being-in-the-world of 'Dasein' is characterised by openness, by a 'having-to-be-open', by a 'standing out to the world', which is how Heidegger read the idea of existence as 'ek-sistence' (see Heidegger, 1962).

One key notion in how Heidegger characterises Dasein's existence is the idea of '*thrownness*' (*Geworfenheit*). We are thrown into the world and we 'find' ourselves there, rather than that we stand before the world (in place and time). It is linked to the idea of 'disposedness' (Befindlichkeit), which has to do with receptivity and even more with the insight that one 'finds' oneself in things that matter to oneself. There is a passive and receptive dimension to this, although with the idea of 'care' (Sorge), Heidegger also suggests that there is an 'attitude' (my own term) at stake here. In the *Letter on Humanism*

Heidegger not only claims that language is 'the house of being' but also that human beings are—and ought to be, so I wish to add—the 'guardians of this home' (Heidegger, 1993, p. 239).

The foregoing indicates that both Husserl and Heidegger are after a very different 'configuration' than the 'setup' of perspectivism. Whereas Husserl goes back to the things themselves, Heidegger doesn't deny things, so to speak, but is interested in something more, namely the way in which things *exist*—the being of things. Yet the configurations they are exploring are not only about how the world 'appears' but also about how the self 'shows up'. For Husserl this is the question how things appear to consciousness or, with his term, intuition. While this is not a matter of consciousness or intuition having a perspective on things, Husserl does rely on the assumption that consciousness is already 'there', that is, before things appear to it. Heidegger takes a different 'route' by thematising the self as 'Dasein' and highlighting qualities such as passivity, receptivity, thrownness, care and guardianship. In doing so, he reveals how 'Dasein' stands out to the world, rather than having its unique gaze on the world from its particular standpoint.

One question this raises, however, is how 'Dasein' might be able to judge what to be receptive to, what to 'let in', what to take care of, and where and when to say no. After all, if there is only resonance—to use a fashionable concept—but no discernment, 'Dasein' just will just become an echo-chamber for anything that forces itself upon 'Dasein', whether it will be world peace or fascism, to name but two options. In *Being and Time* Heidegger discusses this matter in relation to the distinction between authenticity and inauthenticity (see Heidegger, 1962; see also Carman, 2003 and, for the bigger political issues at stake here Gordon, 2010). It is here that another author who stands in the phenomenological tradition is important; this is Emmanuel Levinas.

Levinas's work can be seen as a profound critique of what can be called the 'egological'[5] worldview, that is, the suggestion that the ego would come before the world, both in space and in time.[6] When Levinas suggests that we should see 'ethics as first philosophy' (Levinas, 1986; see also Peperzak, 1995), he is not suggesting that everyone should be responsible and even less so that our ability to be responsible is in some way primary, but rather that responsibility is 'the essential, primary and fundamental structure of subjectivity' (Levinas, 1985, p.95). While this suggestion may sound rather totalising or even ontological, there is an extremely helpful sentence from Zygmunt Bauman which, in my opinion, captures what Levinas is after extremely well. As Bauman puts it, ethics as first philosophy means that 'responsibility is the first *reality* of the self' (Bauman, 1993, p. 13; emphasis added).

What is captured in this phrase is the idea that my self becomes an issue for me—not for anyone or for others but for *me*—when I encounter responsibility. Viewed in this way, we can say that responsibility is 'objective'—we simply encounter it. Yet it is in this encounter that I meet my own freedom, because when I encounter responsibility, when I encounter a question or request that is directed at me, it is for me to figure out whether I say 'yes' or 'no', whether I 'take up' this responsibility or walk away from it. Others can help me, encourage me, distract me, but it is up to me to decide what to do with my freedom and hence with my 'self'. It is, therefore, in the encounter with responsibility that my self begins to matter or, in Heideggerian language, that I am drawn into the world and drawn into my own existence.

Both Heidegger and Levinas thus highlight that the self is not its own origin, does not pull itself into the world, and particularly doesn't assert itself into the world through his or her own perspective on and interpretation of the world. While there is a risk that Heidegger makes the subject too receptive, so to speak, too open to the world, Levinas points at the precise 'moment' where the self becomes an issue for itself as a moment of decision, a moment to say 'yes' or 'no'. The first reality of the self is therefore a moment of heteronomy, but Levinas emphasises that this is 'a heteronomy which does not involve servitude, a receptive ear which still retains its reason, an obedience which does not alienate the person listening' (Levinas, 1989, p. 207). It is, in other words, *not* an argument for saying that we should simply receive everything that comes to us from 'elsewhere'. Each encounter with responsibility is therefore a kind of 'wake-up call' (see also Biesta, 2022a).

Allowing Oneself to Be Found: Taking Things as Calls

The last author I wish to discuss in my exploration of alternative 'configurations' is Jean-Luc Marion. The main theme of Marion's work is 'givenness' and the question he has been pursuing throughout his career is whether and how we can 'make sense' of the phenomenon of 'givenness', bearing in mind that the very point of 'givenness' is that it is given and therefore precisely *not* the product of our sense making or interpretation, which means that it is more accurate to say that Marion is exploring the givenness of phenomena. One of the shortest formulations Marion provides of this is with the 'principle' 'that everything that shows itself must first give itself' (Marion, 2011, p. 19).

This phrase already contains an important point, because it suggests that before any intentional act of knowing, or understanding or sense making can take place, or even before anything can be experienced at all, something must have given itself (see also Roth, 2011). Marion emphasises that his principle does not articulate an interest in *what* is given but in the *how* of 'givenness'. Marion is interested, in other words, in 'givenness as *a mode of phenomenality*, as the *how* or *manner* of the phenomenon', as he puts it (Marion, 2011, p. 19; emphasis in original). This is not, therefore, about 'the immediate given, the perceptive content, or the lived experience of consciousness—in short, of something that is given, but instead of the style of its phenomenalization *insofar as* it is given' (Marion, 2011, p. 19; emphasis in original). Marion therefore is after a *phenomenology* of the phenomenon of givenness, not an ontology or metaphysics (see also Marion, 2011, p. 20).

Marion locates his endeavours firmly within the history of phenomenology, because 'in its most radical ambitions, philosophy, in the form of phenomenology, has had no other goal than this one: to allow the phenomenon to broaden out itself in itself, and to show itself from itself' (Marion, 2016, p. 48). Marion argues, however, that to really fulfil this ambition,phenomenology is in need of a third reduction; not just a reduction to the *thing* (Husserl) or to be *being* of the thing (Heidegger), but to givenness *itself*. About Husserl, Marion writes that he 'sees givenness' and 'sees that one can only reach givenness via reduction, but to a large extent he stays within the presupposition that the given is a matter of objectness, of a theory of the object' (Marion, 2017, p. 77). Marion thus argues that Husserl 'stopped' at the object, just as Heidegger 'stopped' at the being of the object. For Marion, however, 'objectness (Husserl) and beingness (Heidegger) only offer specific and possible cases, but surely not the most legitimate ones, of the naming of givenness' (Marion, 2017, p. 78). Hence the need for 'a third, more original reduction ... namely the reduction to givenness' (Marion, 2017, p. 79).

For the critique of perspectivism I am pursuing in this chapter, it is particularly Marion's discussion about what he refers to as two 'attitudes' towards things, that is a helpful 'operationalisation' of this work on givenness. The first attitude Marion discusses is 'the most widespread' and also 'the one for which we are trained' (Marion, 2017, p. 83). It consists in 'reducing the chances that those [things] around us will surprise us; consequently we continually learn how to better control them' (Marion, 2017, p. 83). In this attitude 'we really count on being able to anticipate situations and accidents, to be able to react, to control, to correct, to secure' (Marion, 2017, p. 83). This is a world where we find ourselves surrounded by objects, 'which, being essentially functional, function because they are intended and conceived to function to our advantage' (Marion, 2017, p. 83). This is done 'so that we are in the centre' (Marion, 2017, p. 83). The first attitude can therefore be characterised as perspectivism.

Through this attitude 'we live in a world that we organize such that we retain from it only those things that can be constituted as objects, only what we can grasp with our intelligibility, under the control of a quasi-master and possessor of nature' (Marion, 2017, p. 83). We do this, we constitute such an 'object-ive' world in order to rule out danger. Yet, so Marion asks, 'what does ruling out danger mean if not keeping away from the unexpected', from that 'which cannot be constituted as object, that against which one cannot protect oneself' (Marion, 2017, p. 83)? While the first attitude is therefore very useful for thinking ahead, that is, for what can be anticipated, Marion argues that 'this rationality ... does not want any of the rest [but] only retains this layer of reality that one can call the object' (Marion, 2017, p. 84).

Marion argues, however, that the object only offers 'a very thin and superficial layer of things. It leaves to the side ... everything it cannot foresee, everything it cannot anticipate, what is said to be unknowable' (Marion,

2017, p. 84). Yet it is precisely here 'that the given is displayed because it characterizes what among things *resists* objectification and is given by its own initiative' (Marion, 2017, p. 84; emphasis added). Marion adds that it is not for us to look for the given, so to speak. Rather, 'in the given, in the phenomenon inasmuch as it gives itself according to its character as nonobject ... a place and moment are described where the ego must know how to allow itself to be found and which it does not decide' (Marion, 2017, p. 85). Here the ego leaves its central position, 'obeys the event, and sees without foreseeing', as Marion puts it, and this 'seeing without foreseeing' is precisely the opposite of the seeing-of-objects (see Marion, 2017, p. 85).

Thus, we encounter a reversal of our 'objective' or 'perspectival' relation to the world. It is not we who command the world. Rather, 'in the case of the given, we find ourselves commanded by the thing, summoned to come experience it' (Marion, 2017, p. 85). We could also say that we are surprised, in the literal sense of 'being seized'. Marion gives the example of a painting in a church in Rome which has a secret point where one must be situated to see the painting. This point, Marion explains, 'is determined by the painting and not by the spectator' so that 'the spectator must obey the painting in order to see it' (Marion, 2017, pp. 84–85).

Although 'givenness' so conceived may give the suggestion that it requires passive receptivity from the side of the ego, to use Marion's term, he stresses that the term 'passivity' is not good enough, precisely 'because I cannot remain passive in front of the event: I make myself available or I avoid it, I take a risk or I run away, in short, I still decide, and I respond even by refusing to respond' (Marion, 2017, p. 85). This means that in order to 'become passive' in such an encounter, 'a certain kind of activity is required; one must leave oneself exposed to things with a certain amount of courage' (Marion, 2017, pp. 85–86) or, in educational language, one must *let oneself be taught* (see Biesta, 2017a, 2017b).

All this, then, opens up 'a different regime of phenomenality imposed on a different regime of subjectivity', as Marion puts it. Here the subject is no longer *before* the world, *before* the phenomenon—in space and in time—but rather 'receives itself from what it receives' (Marion, 2017, p. 86). The word that Marion uses for this different 'stature' of subjectivity, this different 'stature' of the 'I' is, in French, the *adonné*, which is usually translated into English as the 'gifted'.

All this, then, is not just a reflection on givenness and on how the world gives itself rather than that it is 'taken' by us. Marion's reflections also provide an account of how the self is 'returned', so to speak. With regard to this Marion writes, quite beautifully in my view, that 'everything takes on a different meaning if what happens to me is given to me from elsewhere' (Marion, 2017, p. 38). After all, if what happens to me 'is a duplication and a product of myself, then even the most marvellous things lose their meaning' (Marion, 2017, p. 38). But if one instead thinks the world as 'essentially an experience of heteronomy, in other words as election, then, everything is worth being lived, being expected, being desired, everything merits making an effort on its behalf' (Marion, 2017, p. 38)

Marion is fully aware however that before trying to respond to the call 'there is the more difficult thing, which is … to discover that there is a call, that is to say, being able to interpret what is as what comes to us' (Marion, 2017, p. 39). And 'this decision to take things as calls … decides everything else' (Marion, 2017, p. 39).

Concluding Comments: A Change of Heart

In this chapter I have suggested that the idea that the world—natural and social—is simply there for us to have a perspective on or view about, is not just a problematic idea, but also a very specific 'configuration', that has emerged at a particular point in time which Heidegger characterises as the age of the world view of world picture. The 'arrival' of perspectivism is, in a sense, a gain, as it brings about a field in which the individual obtains a defined position or standpoint from which the world is observed. I have characterised this gain as freedom and, more specifically, as the freedom of having a standpoint and the freedom of having a perspective. This is a gain compared to non-perspectival setups in which the individual can be anywhere and thus ends up nowhere or, to be more precise, nowhere in particular.

Yet this particular gain, which can be seen as a modern gain, also comes with losses. One loss, which is relevant for the overall theme of this book, is the loss of the world—both the natural world and the social world—because in the age of the world view the world turns into an object for observation and is, in this sense, objectified. And this objectification, so we might say, is the first step towards the maltreatment of the world, the consequences of which we are witnessing today. We could say that as a result of the objectification of the world, the world loses its voice or, with a notion that is quite close to that of voice: the world loses its authority in that it no longer appears as 'author', as an entity that has something to say to us from itself. This is what happens when the world becomes picture and the individual becomes world-viewer or perspective-taker.

The point I've tried to make in this chapter, is that the problem we are facing does not lie in a particular perspective on the world but in the very idea that our relationship with the world is perspectival, that is, is one of us having a perspective on the world. It is not, therefore, that we should get rid of particular perspectives and replace them with better ones—such as swapping a Western perspective for an indigenous one, or a colonial perspective for a postcolonial, or monological for a relational, or a humanistic for a posthuman one. It is rather that we should try to overcome the logic of perspectivism *itself*.

With regard to the world this requires what, in a quite literal sense, we might call a re-animation of the world, as it could be argued that the objectification of the world has turned it mute and soulless. With regard to ourselves, it has, perhaps, something to do with overcoming a kind of arrogance entailed in perspectivism, as if the world is waiting for our opinion, and wouldn't be able to carry on without it. The simple 'turn' that is required, which could also be the most difficult turn to make, is 'this decision to take things as calls', as Marion has put it. It is interesting that Marion refers to this as a *decision*, because this may well sound as a form of perspectivism, as if it were up to us to decide whether the world is calling us or not. Yet I do not think that this is what Marion is trying to say. The decision, so it seems to me, is not a decision about the world, but it is *a decision about my own life*. It is the decision to let the call of the world 'in', to not deny it, to conduct my life, from now onwards, out of this 'attitude'. A change of heart, as I have called it in the beginning of the chapter.

There is much that can be said about education in relation to all this, but in a sense it is also very simple because in its most basic—which also means most fundamental—form, education is nothing more, but also nothing less, than the art of redirecting someone else's attention (on this particular definition see Benner, 2020, p. 21). Here lies the critical difference between learning—which is about our own attention—and being taught—which is about our 'own' attention being *re*directed onto something. This is precisely what, by definition, we cannot do ourselves (we can only *direct* our attention) but for which an educator or teacher is needed. Education is, in other words, a matter of pointing, that is, of trying to 'catch' and turn someone's attention through a gesture of 'look there' or, more precisely, 'attend to that' (on this see Prange, 2012a, 2012b; Biesta, 2022b).

There is, of course, always the possibility that the student's attention is immediately 'filled' with interpretations and perspectives, which happens when education operates within the logic of perspectivism in which students are urged to opt for particular perspectives or, in more constructivist approaches, are urged to generate their *own* perspectives on the world. Although the pull of educational perspectivism is strong—and according to many is the only conceivable option for education—the upshot of what I have presented in this chapter is the challenge, first of all for teachers, to keep students away from the pull of perspectivism. It is the challenge to keep pedagogy sufficiently 'poor', so to speak (on this see Masschelein, 2010), so that the 'address' of the world can arrive, so that the world can be encountered as appeal, as question, and perhaps even as demand.

Rather than trying to change students' perspectives—replacing 'wrong' perspectives with 'right' perspectives—which would always tie students to the logic of perspectivism, this suggests an altogether different 'avenue'. And while it is too presumptuous to see this as a full-scale solution to the environmental crisis we are in, it may at the very least provide a very different and, in my view-which-is-no-longer-a-view, also more hopeful 'opening'.

Notes

1 Readers can find more educational detail in Biesta (2017a, 2017b, 2021), albeit that I now regret having used the phrase 'view for the present' in the subtitle of the latter book. I also highly recommend Roth (2011).
2 The German word translated here as 'man' is 'Mensch', which would be better translated as 'human being'.
3 Heidegger makes the point that as soon as Christianity is seen as a world view it 'adapts itself to modern times' (Heidegger, 1976, p. 342), which can be understood as the suggestion that as soon as one speaks about a Christian worldview one has already secularised Christianity. On this issue and its implications for religious education see also Biesta and Hannam (2023).
4 Donald Trump is a sad but also potentially dangerous example of such a denial of reality.
5 For the term 'egology' see Levinas (1969, p. 35).
6 Following what I have said so far, this also means that any worldview is actually egological in this sense.

References

Bauman, Z. (1993) *Postmodern ethics.* Oxford: Wiley-Blackwell.

Benner, D. (2020) *Umriss der allgemeinen Wissenschaftsdidaktiek.* Weinheim: Beltz/ Juventa.

Biesta, G. (2017a) *The rediscovery of teaching.* New York: Routledge.

Biesta, G. (2017b) *Letting art teach: Art education after Joseph Beuys.* Arnhem: ArtEZ Press.

Biesta, G. (2021) *World-centred education: A view for the present.* New York: Routledge.

Biesta, G. (2022a) Have we been paying attention? Educational anaesthetics in a time of crises, *Educational Philosophy and Theory*, 54 (3), pp. 221–223.

Biesta, G. (2022b) Why the form of teaching matters: defending the integrity of education and of the work of teachers beyond agendas and good intentions, *Revista de Educación* 395, January–March, pp. 13–33.

Biesta, G. & Hannam, P. (2023) The world is not enough: Religious education beyond worldview perspectivism, in L. P. Barnes (ed), *Religion and worldviews: The triumph of the secular in religious education.* New York: Routledge, pp. 99–117.

Carman, T. (2003) *Heidegger's analytic: Interpretation, discourse, and authenticity in 'Being and Time'.* Cambridge: Cambridge University Press.

Carman, T. (2006) The principle of phenomenology, in C.B. Guignon (ed), *The Cambridge companion to Heidegger*, 2nd ed. Cambridge: Cambridge University Press.

Edgerton, S. Y. (2009) *The mirror, the window, and the telescope: How renaissance linear perspective changed our vision of the university.* Ithaca: Cornell University Press.

Gordon, P. (2010) *Continental divide: Heidegger, Cassirer, Davos.* Cambridge, MA: Harvard University Press.

Gruwez, C. (2009) *Zeitgenosse werden: Ein manichäischer Übungsweg.* Dornach: Pforte Verlag.

Heidegger, M. (1962) *Being and time*, trans. J. Macquarrie and E. Robinson. Oxford: Basil Blackwell.

Heidegger, M. (1976) The age of the world view, *Boundary 2* 4(2), 340–355.

Heidegger, M. (1982) *The basic problems of phenomenology*, trans. A. Hofstadter. Bloomington: Indiana University Press.

Heidegger, M. (1993) Letter on humanism, in D. F. Krell (Ed) *Martin Heidegger: Basic writings*, revised and expanded edition. London: Routledge, pp. 217–265.

Heidegger, M. (2002) The age of the world picture, in M. Heidegger, *Off the beaten track*, ed., trans. J. Young and K. Haynes. Cambridge: Cambridge University Press, pp. 57–85.

Husserl, E. (2001) *Logical investigations*, 2nd ed. 2 vols. London: Routledge.

Levinas, E. (1969) *Totality and infinity: An essay on exteriority*. Pittsburgh, PA: Duquesne University Press.

Levinas, E. (1985) *Ethics and infinity: Conversations with Philippe Nemo*. Pittsburgh, PA: Duquesne University Press.

Levinas, E. (1986) *L 'éthique comme philosophie première. Actes du colloque de Cerisy-la-Salle*, 23 August–2 September.

Levinas, E. (1989) Revelation in the Jewish tradition, in S. Hand (ed), *The Levinas reader*. Oxford: Blackwell, pp. 190–211.

Lingis, A. (2007) *The first person singular*. Evanston, IL: Northwestern University Press.

Marion, J.-L. (2011) *The reason of the gift*. Charlottesville, VA: University of Virginia Press.

Marion, J.-L. (2016) *Givenness and revelation*. Oxford: Oxford University Press.

Marion, J.-L. (2017) *The rigor of things: Conversations with Dan Arbib*. New York: Fordham University Press.

Masschelein, J. (2010) E-ducating the gaze: the idea of a poor pedagogy, *Ethics and Education*, 5 (1), pp. 43–53.

Peperzak, A. (ed.) (1995) *Ethics as first philosophy: The significance of Emmanuel Levinas for philosophy, literature and religion*. New York: Routledge.

Prange, K. (2012a) *Die Zeigestruktur der Erziehung. 2. Auglage. korrigiert und erweitert*. Paderborn: Ferdinand Schöningh.

Prange, K. (2012b) *Erziehung als Handwerk. Studien zur Zeigestruktur der Erziehung*. Paderborn: Ferdinand Schöningh.

Roth, W-M. (2011) *Passibility. At the limits of the constructivist metaphor*. Dordrecht: Springer.

3

WHEN THE PLANTS TALK BACK!

Teaching in Response to a Call from Elsewhere

Carl Anders Säfström

Introduction

In this chapter, the relation to plants (nature) is viewed from our plant-ness. It gives examples of an alternative cosmology, in which language is not separated from and mirroring the rationality of nature (Rorty, 1980), but rather in which place and language are merged into one wholeness and, as such, displacing fundamental distinctions of modern forms of life (Fish, 1989; Säfström, 2023b). Furthermore, two profound ecological experiences of such plant-ness are explored, and therefore also make problematic the fundamental distinction emerging from Aristotle in which men of lower standing, women, children, and enslaved people were considered to be like plants, inhuman. A consequence of the sharp split between man and plant in Platonic–Aristotelian orthodoxy and modern capitalistic forms of life is understanding what nature is in man to be controlled and or used. In this respect, they are complementary forces from the point of nature; they have similar consequences, and the Platonic–Aristotelian orthodoxy creates the reality in which the use of and control of nature make sense. It makes a capitalist life possible, in which nature is only a resource for man to use as he pleases. Both Aristotle and later Immanuel Kant deal with this issue and, in complementary ways, establish education as that force which would transform the inner evil of man, his nature or what I call 'plant-ness', to complete, for Kant, God's plan of perfection (see further Cassin, 2014, pp. 153–154; Säfström, 2023b, pp. 31–32). For Plato (Bloom, 1991), the role of education was to perfect man to perfect the state, making education and teaching into fundamental tools for producing a particular state of perfection, producing man himself as internally linked to the state. For a man to realise himself, he simultaneously realises the state (see further Jaeger, 1986 [1944], p. 259;

DOI: 10.4324/9781003456636-4

Säfström, 2023a). The patterns the subject and object established in such a state mirror each other perfectly. *Perfection*, in other words, is socially, politically, and ethically dominating the educational state[1] and the man taking shape within such state as a perfect object for the will of the state.[2]

The following will resist education as a significant force of domination and objectification of 'man' from the point of the state and instead defend the plant-ness inscribed in humanity. The form of resistance developed is inspired by the tactic of 'talking back' (hooks, 2014) to resist dominant forms of oppression:

> Moving from silence into speech is for the oppressed, the colonised, the exploited, and those who stand and struggle side by side. It is a gesture of defiance that heals and makes new life and growth possible. It is that act of speech, of 'talking back,' that is no mere gesture of empty words, that is the expression of our movement from object to subject–the liberated voice.
> *(hooks, 2014, p. 9)*

To teach from elsewhere is to talk back with a liberated voice, to open us to the present in which life is shared with all living and, for some, with trees and stones (Andreotti, 2021). Through the stories told in the following, a view from elsewhere is established, and the chapter ends with teaching as a response to the call from the plant. To be clear, Aristotle's description of the sophists, who were men of lower standing and 'women, children, animals and slaves' (Cassin, 2014, p. 6) as *plants*, are not only degrading and excluding those people from mattering in the educational state but also degrading the actual plant. So, in this text, 'the plant' is a name for actual plants and those excluded by default of a Platonic–Aristotelian orthodoxy; it symbolises what nature is in humans and nature itself.

The stories of such plant-ness are stories about the interconnection between man and nature and rely on what Rancière (2007b) calls *le partage du sensible*: Stories, as part of everyday life, always touch the sensible realm simultaneously one's own and shared with other sensing beings. That is what *le partage* means and what makes it into a story. The story allows us to touch and be touched as sensing beings.

The chapter explores what will follow for education and teaching if we teach beyond the fundamental distinction between man and plant and if we teach from elsewhere. Rather than confirming teaching as an expression of inequalities at work, the exercise of a master over his or her student, the following practice education and teaching as events intervening in the reproduction of social order and nature as mutually exclusive expressions of who and what is dominating/being dominated, who and what is humane/inhumane. Rather than reproducing the relation between past and future in terms of sameness (be your master) and the relation between man and state regarding natural inequality of a fixed hierarchical order (Rancière, 1991,

2007a), teaching in this chapter concerns the equality of the unharmonisable plant. That is, concerns 'the plant' that talks back! Such teaching is performing the heterodoxy of a sophistical practice (Cassin, 2014; Säfström, 2023a), resisting humanism to serve humankind. It is an active nonviolent resistance in which the border between who and what is 'grieveable' (Butler, 2020), what is ethically significant, is extended to include plants and all living. Teaching from elsewhere, the chapter will conclude that it is loyal to the pluralism of humankind by performing the equality of unharmonisable plants. It is teaching that lands among the heterogenic multiplicity of people, the multitude of animals and plants, and in an environment on which we all depend for survival but which is cut off from concerning us by modern forms of life (Andreotti, 2021).

The following steps will be taken in the chapter: I will first make problematic the idea of perfection as an end and means for education, followed by a section in which the imperfection of the plant is embraced rather than repressed, and after that comes a section in which two stories of deep ecological experience are told. The last section summarises the issues dealt with in the paper. It discusses teaching as a tactic from elsewhere, intervening in oppressive authoritarian forms of life, establishing a plurality of ways of life beyond the drive for perfection of modern life, and beyond the tendency of capitalism in late modern times to reduce all variation to its own 'truth' (Mouffe, 2005; Rancière, 2007a; Berardi, 2017; Andreotti, 2021).

Decentring Perfection as the Aim and Goal of Education

In capitalism and nationalism, both (Orelllana & Michelsen, 2019) the male body and what is culturally assigned to it, strength, aggression, and competition, are the driving force of the economy as well as an image of the state itself, as one organism in which the inner psychological patterns of being a man are compatible with the socio-political patterns of the perfect state (Jaeger, 1986 [1944], p. 259; Berardi, 2017; Butler, 2020). Man and state mirror each other in perfect symmetry, and the role of education and teaching is to hold up that mirror to secure order and stability and reflect what is considered to be the real as such (Rorty, 1980). That is, such 'mirrors' are not only to create a metaphysical image/ideal of a perfect state of things but to materialise a particular order of things as perfect reality. Central for culture in late capitalism then is a specific form of perfection, in which the experiences of loss and grievances (Säfström, 2021), following from the climate emergency we live in as well as the destruction of social equality as aim and goal for liberal democracies, following neo-liberalism (Butler, 2015), take *violent* forms (Berardi, 2017). It mobilises a drive for a violent resurrection of the male body with devastating consequences (Berardi, 2017), feeding the turn to fascism and ultra-nationalism in many European countries as well as globally and establishing authoritarian capitalism as hegemonic (Giroux, 2018).

The ideal of perfection and the aim for a perfect state of things embedded in authoritarian capitalism and nationalism are ultimately based on Platonic politics and Aristotelian science, in that both locate education as that tool for which perfection of man and state is the sole aim and function of education (Jaeger, 1986 [1944], p. 259). The educated, perfect body of man is simultaneously an expression of the ideal body of the state. For Plato, the state is one organism comparable with a body, which makes it impossible for a pluralism which is not already part of this One perfect or perfectible body (Cassin, 2014; Rancière, 2007a; Arendt, 2005). Such anticipations of 'the perfect One' concern not only the strict relation between state and bodies but also define the totality of the reality in which those bodies are to appear and form. Moreover, as a Platonic–Aristotelian line of thought informs educational thought, it carries this 'reality' necessary for making sense of and within education. At the heart of this type of sense-making 'reality' is an idea of the perfection of one and undivided reality for those who can be included in such reality. It is a *perfect* reality only certain people carry to express their being. They embody, or have the capacity to incorporate, the natural state of things; they live the truth of perfection. For Plato and Aristotle, it was the aristocrats who were the only ones to be educated, to be able to embody *aretē*, to incorporate culture, and to be perfected through education in a hierarchical relation of masters and mastered (see further Säfström, 2023a; Rorty, 1980; Cosigny, 2001; Rancière, 1991; Jaeger, 1965 [1939], 1986 [1944]; Cassin, 2014).

Modern forms of education generally are, to a large degree, still captured by this image of perfection by domination, the need to be perfect certain people. In contrast, others are *imperfectible* as such. Be it because of a perceived lack of intelligence, genetic predispositions, social and political exclusions, gender, sexuality, ethnicity, or what have you. Instead, imperfections are distinctly part of everyday life, in which people live in imperfect circumstances and environments in which they depend. For the state of perfection, the consequences that are anticipated as following from the imperfectness of everyday life are the threat of ambiguity, chaos, and despair, producing what Bernstein (1983, p. 18) calls the 'Cartesian angst' and driven by such angst and fear the ones deemed imperfect needs to be corrected dealt with, and in worst case dismissed and annihilated (Bauman, 1991). That is, imperfection is defined from a particular vantage point, that of a perceived perfect point of power from which meaning flows. That is, and again, imperfection is only such from a certain point, formed and made meaningful in its cosmology in its reality, and in this reality, driving progress as a constant unfolding of its superiority. Alternatively, in other words, as Jacques Rancière (1991) has pointed out, modern forms of education are, to a large degree, themselves expressions of the very drive of modernity to progress towards an end goal of perfection for specific people rather than others.

Time itself in those forms of education/modernity tends to be teleological, and pedagogy, consequently, can only be understood as *the managerial function of time* for progression towards an ultimate goal, which tends to be a perfect being and a perfected state (all in an ideal state of mind).

The idea of perfection in modern forms of education is what unities images of the state, science, and God (Cassin, 2014). Immanuel Kant (2012) gave education and pedagogy the role of perfecting God's work on the imperfect nature.[3] Consequently, for Kant, pedagogy has a moral obligation to redirect the child from its inner nature, its imperfections, towards a perfect moral life in line with God's plan (see Cassin, 2014, p. 154), which is always a perfect life in a particular idea of a state of a certain kind. According to Kant, there is, therefore, a political, scientific, and moral obligation of man to perfect nature, himself perfected as he goes about his divine task.

Consequently, politics, science, and education, in a Platonic–Aristotelian line of thought, and here I include Kantian education, are all forces to overcome imperfection and all that is imperfect: nature, everyday life, the child, and more. The foundational *orthodoxy* of Platonic–Aristotelian worldviews establishes the distinction and definition of imperfect life as absolutely separate from perfect life and as the fundament of our Western/European culture. Such fundament is based on the distinction between humans and the inhuman nature, in which 'women, children, animals and slaves' (Cassin, 2014, p. 6) were all, according to Aristotle, again, to be considered as *plants*, as an expression of imperfect nature. Therefore, the plant symbolises all that is regarded as imperfect, not to be perfected but as *incapable* of being perfect within such orthodoxy.

In Platonic–Aristotelian informed educational thought, only the aristocratic child can be perfected, or to be precise, only the male child can be perfected and is already considered perfect. The ultimate aim of education and teaching is to manifest its perfection while establishing the imperfect as such. The managerial function of pedagogy in modern education in this tradition of thought is to execute this fundamental distinction between what and who is possible to perfect and what and who are, by definition, not likely to perfect, being fundamentally and eternally imperfect. Schooling identifies and separates the perfect from the imperfect ones in modern life.

Man's plant-ness is a disturbance, an irritation, from within himself. My plant-ness, my imperfection, in modern life cannot be harmonised with the idea of a perfect state and the authoritarian capitalism we live in today. Therefore, in the following, I embrace my imperfection and plant-ness as an act of resistance and talk back from such a position. I will 'stop, look, and listen' as Dewey (1966) said; I will be looking and listening elsewhere. In the following section, I will recall a language that does not operate on the assumptions of sharp distinctions between perfection and imperfection, between man and nature, or between the actual plant and ourselves. I do so to take on the imperfection of the dominant discourse and turn it onto itself as a counter-tactic of the unharmonisable plant talking back!

Talking Back: Living the Imperfect Life of Plants

A Sámi daughter told me[4] at a symposium in south Norway that the Sámi language (Sápmi) does not work through categorisations like Norwegian, English or other modern languages. The present reality is instead expressed in and through the poetry of words, laying bare its surroundings, not abstracting, categorising, distancing oneself from the lived embodied life, but rather touching trees, grass, mountains, and reindeer, all as experienced as part of the concrete and spiritual world one inhabits in and through the poetry of the language. No separation. Man, women, children, animals, and plants are all equally present in the language as they are not culturally constructed out of an ideal of perfection, in which the plant-ness of life equalises radical evil to be controlled from a centre. That does not mean that social hierarchies do not exist today, particularly after colonial powers infiltrated the language, or that power structures only was imposed on Sámi culture, but that Sámi language embodies another reality in which the world takes shape differently and in a way in which the plant-ness of life is part of that reality. The structure of the language is a consequence of its use in everyday life. You live *with* people, animals, and plants as a *practice* of telling stories, and it is, as such, already giving rise to cosmology that is different from modern languages. Gro also tells me that since the language is emerging as a practice of storytelling and poetry, it does not have the same structure as stories in Norwegian or English. It does not have a beginning-middle-ending structure, with a clear conclusion, but is expressed in the poetic way a story unfolds and for the listener to draw his or her conclusions. The language is, therefore, open-ended and involves the other already in its composition; it is non-hierarchical and inclusive. Therefore, what is to be understood as reality or nature is, rather than described from a distance, unfolding as a consequence of the poetic expressions themselves in the here and now of the place itself. Land (nature) and language are not separated but intra-dependent and, therefore, part of the same concrete and spiritual cosmology in which one lives a Sápmi life.

Therefore, the land issue, the expansion of Capitalist culture into the heart of Sápmi, in which, ironically, the breaking of the land is said to save the earth from pollution by mining materials for batteries to feed cars and eventually planes, is also breaking into the heart and soul of Sápmi as such. Sharon Todd (2023) writes, concerning the essential difference of worldviews at display:

> The mine will disrupt traditional reindeer breeding tracks thus threatening not only the livelihood of those Sámi who herd them, but the future of the reindeer themselves. The British-owned mining company offered statistics in a supposed gesture of consolation: the mine would represent just 0.5% of the available pasture lands of 1.019 sq miles. In response, Jonas Vannar, a herder, stated: 'You can show anything with statistics. If you compare it to someone shooting an arrow to your heart, it's a pretty small hole when

> you calculate the area of the wound compared to the entire body. Then it's surely not possible to die from an arrow in your heart?'
>
> *(Todd, 2023, p. 1)*

Todd refers to Bruno Latour (2017) and clarifies the point that 'this is not an effect of simply seeing 'the world' from different, conflicting vantage points; rather, it is that we indeed inhabit different worlds' (Todd, 2023, p. 1). We see the world of capitalist culture breaking into the heart and soul of Sámi life, the latter in which 'the world is a place of profound interconnection' with other living beings and non-living entities, says Todd.

The world in which a capitalist life takes form is constituted by *violence and domination* over the everyday life of people, their language practices, and the land they live (Butler, 2020). The capitalist world is not simply 'a world' as a noun. However, instead, a verb, a 'worldling' (Todd, 2023), that is, in my words, a performative cosmology of capitalist worldling, the making of *a* world: Capitalism is continuous performativity of reduction to its cosmology to One singularity. As such, it is alien to cosmologies that would unfold different ecologies of relations between people, animals, plants, and non-living entities. However, those radical heterodoxies traditions of 'languaging' in which the multiplicity of life and its polyvalence (Bauman, 1993) are not to be reduced but are the nonreducible conditions of living among other lives that matter do exist, as shown above, and tap into those means to enter into a radically different plurality of worlds. I have elsewhere been writing about the sophistical practice (Cassin, 2014) of living in the here and now and the consequences that follow in terms of education, democracy and a liveable life (Säfström, 2023a) as one expression of a profound tradition of heterodoxy thoughts and practices, suppressed, but never wholly erased in intellectual history. The exciting thing here is that this Sophists (500 BCE) tradition of thought is based on the practices of telling stories (Cosigny, 2001), the practices of living well with a multitude of others recognised in their otherness (Cassin, 2014), and the practices of walking in and between different worlds (Jaeger, 1965 [1939], 1986 [1944]), or city-states, as it were for the sophists. Sophistical practice is an embodiment of a profound non-reductive cosmology in which living in the here and now means confronting the issues at hand, for example, the extreme situation of a climate emergency in which we currently live, to find ways of moving on together but not necessarily in the same way as our 'fellow adventures' (Rancière, 1991, p. 64).

Furthermore, for the Sophists, that we are formed by the language and land we live, rather than by the blood we inherit (as for Plato and Aristotle, see Jaeger, 1965 [1939]), opens a profound interconnection with all living and non-living entities. The point here is that despite the tendencies of reduction intrinsic to a capitalist life, cracks and breaks always come to us from the outside 'like that brute way in which stones impinge on our feet' Rorty (1980, p. 375). That is, just because a capitalist worldling feeds a drive of reduction to sameness and, therefore, exclusion of other ways of perceiving and living the world does not mean it always succeeds. There is always a possibility, as Jacques Rancière (1991, p. 90) says, to refrain from 'falling into the land of inequality' and the social organisation necessary for capitalism to work, as well as refraining from living in the overall ideological monoculture of capitalism. Such life may not be perfect, as if any life could be, but it would be open to other ways of connecting to the world, of living with and among multiplicities of cosmologies and ecologies.

There is a sense in which profound challenges to our capitalist world happen beyond willingness or choice but instead suddenly break into our self-understanding and connect us to the land where we are living and to the complexity of interconnected ecologies of relationality, expanding over and beyond the humankind/inhuman/plant/ divide. To explore such an event, I will use the following telling two stories, which both, even if in slightly different ways, speak to what philosophers have called 'deep ecological' experience (Næss, 1986).[5]

Two Stories about Trees and Deep Ecological Experiences

Where I grew up, in a suburb of an industrial town in Sweden, on our street right opposite our apartment building, was a 300-year-old oak tree standing. As such, the newly built neighbourhood was carved out of a forest with plentiful oak trees. The old oak on our street was loved by children and adults alike; it gave the entrance to the apartment buildings a certain flair but also connected the neighbourhood to its past and to the present land on which we spent our lives. The city decided, without any consultation with people living in the area, that the old oak tree was to be cut down. Furthermore, one day, we woke up to the fact that the tree was gone, had fallen, and when I came home from school, it had already been taken away. Left was an impressively large stump and a group of people standing around, in shock and mourning, staring at the stump. Over the coming days, people from the neighbourhood would come and pay their tribute to the dead tree, to talk about how beautiful it had been, how they or their children had climbed the tree and how it had had an impact on their connectedness to the place, to the land, to other living beings in the neighbourhood, as a sign of our interconnected lives there and then. There was no separation between the tree and the community's people in that moment of grief. To kill the tree was affecting something precious for the whole community; it was not 'just a tree'; it was a living life and all life 'worthy of value' (Butler, 2020, p. 24). The old oak had deeply impacted our lives, and it was grieved when taken from us. It had a worth attached to it; the tree was a valuable life among other lives and, as such, equally grievable when lost (Butler, 2020). In that moment of grief, there was no absolute separation between different life forms; it was instantiating itself into the hierarchisation between human life and the life of nature necessary for capitalist culture to work. There was no 'radical evil'[6] to be controlled through education, but a lived, embodied recognition of a profound equality in life's worth.

Growing up in Sweden, where you are always close to a forest or park, the attachment to trees has followed me throughout my life. As a student in Uppsala, I had what a philosopher I met in Dubrovnik at the Inter-University Centre before the last Balkan war called a deep ecological experience with a pine tree: I had been sick for a week and been sleeping and reading and sleeping and reading some more for the entire week stayed inside the house. I lived in a student collective in an attached house bordering a forest with mixed types of trees, among them a beautiful slant with birch trees and an open area that had been partly cultivated but where some old and impressive pine trees had been left growing tall. Walking through the forest, I came to that open area with the pine trees and suddenly felt a strong urge to hug one of the pines. I did, and immediately felt a deep connectedness with the tree, a strong impression that life was running through our joint veins, that we shared the same life force here and now. It was a deeply felt impression on my whole body and soul simultaneously. We did and do share the force of living, unseparated in that moment. It lasted for some minutes, but the incident changed how I understand the world I live in and through; I reconnected to my plant-ness and a profound insight into life *as it is*, here and now, as the Buddhist philosophers say (Dōgen, 2006). The tree had a worth attached to it, coming from the tree to me, rather than its value in calculating its worth. I was humbled by the directness of its presence as a living being, calling me to respond as another living being, so I hugged the tree. I did not experience this call as a mystery or metaphysics of nature but as a concrete and material spirituality, a present filled with meaning. The interconnectedness decentred the ego, and 'the I' could appear in an ecology of shared interconnections from which a world emerged. A world that was already there was calling me to partake. The call to presence, to the present of here and now of shared life, opens the present for time. Time is 'being-time', shared in its exposure (Dōgen, 2006). The present is not always present, but when we are called to experience the directness of the interconnection of energies of life, time appears as does being itself.

The everyday living *with* and among other people, animals and plants, on which we depend for our very existence, are constantly unfolding worlds interconnected through an 'ecology of encounters' (Todd, 2023) from which we cannot escape. Not without forgetting our plant-ness, but even so, those encounters constantly remind us of our plant-ness.

That we live in and through those ecologies of encounters means that worlds are not separated; there is no diversity of worlds to which we need to find connections from a distance. No exotic experiences of living are essential to certain people, only attached to other forms of living at different places than our own, from which we are excluded and only ever standing outside looking in to learn their way of life. Of course, we can always learn from others, but fundamentally, such displacement of our being risks alienating our plant-ness and always risks othering people, places, and plants while reinstating our centrality. Instead, it is the case that different worlds are inhabited in every subject, that everyone living is interconnected through the experience of living in the here and now, of living being-time. To realise different worlds of our existence is a consequence, as the examples have shown, of an unfolding of the nonreducible multiplicity of being, in presence, an assortment that is always already there. To open ourselves to such multiplicity, to a profound interconnectedness, and to the ecology of encounters simultaneously means resisting the monoculture of capitalism. Resistance is a consequence of living an imperfect life, with and among other plants, in the face of a perfect life of a capitalist ego.

Teaching from Elsewhere

If the above is trying to describe the break of the convention of living a capitalist life as something that happens to us, if we are willing to open ourselves to the call from the plant to hear its scream, it also goes beyond our will. As such, openness is a place, a presence from which teaching emerges. The touch of the plant teaches us. That is, teaching is not always, as Joseph Jacotot (1770–1840) thought, and as Jacques Rancière (1991) has been exploring, about how to direct the will, but also emerges as a consequence of the non-willing to accept what is present regardless of what we want. Furthermore, it is not the case that we need to teach for those profound ecological experiences to happen. We are not transmitting anything in those cases. However, we can make sense of the breaking through of our conventional world in its exposure to the possibility of another world here and now by recognising our plant-ness being in balance with the world of plants. The art of teaching is all about being in balance; it is about softening our membrane, singling us out from being touched by the interconnectedness of all plants, animals and people alike.

The robotics communicator Subriya Rathi posted a recent art installation to LinkedIn in December 2023. The installation was set up to help a plant defend itself armed with a machete. It was done by attaching an industrial robot arm holding a machete directed by electrical impulses from the plant. It made the robot arm swing the machete as if chasing off anyone and anything threatening the plant. The point of the installation, as I read it, was not whether the plant was consciously defending itself but rather to pinpoint the *de facto* drastic need to protect itself from the vast devastation of extinctions humanity is inflicting on anyone or anything else on this planet which is not considered 'worthy of value' (Butler, 2020, p. 24), or whose only value is to be monetised through capitalism and consumed.

Resistance to capitalism's destructive forces and its consequences for the plant involves acknowledging our own weakness, imperfectness, and interconnected lives and emphasising our capacity to live the world *as it is here and now* and, as a consequence, encounter all life. Resistance is not passive but an active, nonviolent struggle to join forces to talk back, to 'recast the living as worthy of value' (Butler, 2020, p. 24).

Suppose we understand teaching, as Rancière (1991 p. 11) does, as linking different worlds and extending those worlds to all living. In that case, the task of teaching is to link different struggles addressing violence against plants and against humans and decentre the very capitalist ego in the centre of our ways of living a capitalist life of violence, domination and consumption. Such teaching is performing the heterodoxy of a 'sophistical practice' (Cassin, 2014; Säfström, 2023a), resisting humanism to serve humankind. It is an active nonviolent resistance in which the border between who and what is 'grieveable' (Butler, 2020) is extended to include plants and all living. Teaching as such is loyal to humankind by performing the equality of the unharmonisable plant. It is teaching that lands among the heterogeneous multitude of people, the assortment of animals and plants, and in an environment on which we all depend for a liveable life.

Teaching is about linking those moments of opening of places and spaces, where we are called upon from 'outside' ourselves, that moment of *kairos*, themselves unrepeatable but profoundly meaningful in teaching us something valuable, adding worth to that moment. Those moments of *kairos* are not predetermined in a *telos* of steps to be taken, where those steps are already meaningful from the point of perfection, in cases where linking always comes down to comparisons regarding better or worse, winning and losing. The *techne* of teaching is instead the very linking of those moments of *poesis*, autotelic and unique as they are expected of all living beings, and for some, extends to the non-living as well. Those moments take shape from within an ecology of encounters, in which man, woman, child, animal and plants are on the same plane, in the same 'mixture', moving together in concert for meaning and being-time to appear. Teaching makes time for being here and now. As such, teaching gives rise to another starting point for resistance, in which equality is not about more of the same but instead of listening and speaking across different worlds as they take shape in the mixture of encounters and through the poesis of languages. It is to be talking back to power in more than one sense since such poesis of languages is also, in consequence, a politics of languages, that is, not through politics as the will to power (Mouffe, 2005; Rancière, 2007a), but politics as the capacity to apprehend other worlds than the singularity of One, and the practice of linking those worlds in ways that makes new beginnings possible.

Notes

1 On the issue of an 'educational state' of modern times, see, for example, Säfström (2020).
2 This is also why Gert Biesta's (2014) suggestion of education as subjectification is immanently political and in line with this chapter's argument. It breaks not only with the Platonic–Aristotelian orthodoxy in educational thought but also with a specific location of education as the prime tool of shaping the citizen from an absolute point of power.
3 I refer to Immanuel Kant here for two reasons; one is that Barbara Cassin (2014) shows that for Aristotle as well as Kant, the question of 'radical evil' is central to their understanding of education, and secondly, because Kant opens for a particularly influential tradition of education, developed further by Johan Friedrich Herbart (1776–1841) (Herbart, 2021) among others. As such, Kantian aspects are, to other possible philosophical traditions, particularly influential in educational thought up to this day, spanning from educational philosophy and theory to curriculum theory and didactics.
4 I recall this conversation with permission from Gro Gardå. The following, though, is my understanding of what Gro told me.
5 The term 'Deep ecology' was established by Arne Næss (1986). While there is a discussion of the usefulness of 'deep ecology' as a political movement, its shortcomings and benefits as a term and field of study, I am here talking about the experiences such ecology tries to capture, what I call 'deep ecological experiences'.
6 A consequence of the sharp split between man and plant is understanding nature regarding radical evil in man. Both Aristotle and Kant deal with this issue and, in complementary ways, establish education as the force that transforms this inner evil in order to complete, for Kant, God's plan of perfection. See further Cassin (2014, pp. 153–154) and Säfström (2023b, pp. 31–32).

References

Andreotti, V. (2021). *Hospicing modernity: Facing Humanities Wrongs and the implications for social activism*. North Atlantic Books.

Arendt, H. (2005). *The promise of politics*. Schoken Books.

Bauman, Z (1993). *Modernity and ambivalence*. Polity Press.

Bauman, Z. (1991). *Modernity and the Holocaust*. Polity Press.

Berardi, F. (2017). *Futurability: The age of impotence and the horizon of possibility*. Verso.

Bernstein (1983). *Beyond objectivism and relativism: Science hermeneutics, and praxis*. Blackwell.

Biesta, G. (2014). *The beautiful risk of education*. Paradigm Publishers.

Bloom, A. (1991). *The Republic of Plato*. Basic Books.

Butler J. (2020). *The force of non-violence: The ethical in the political*. Verso.

Butler. J. (2015). *Notes towards a performative theory of assembly*. Harvard University Press.

Cassin, B. (2014). *Sophistical practice: Towards a consistent relativism*. Fordham University Press.

Cosigny, S. (2001). *Gorgias: Sophist and artist*. University of South Carolina Press.

Dewey, J. (1966). *Democracy and education*. Macmillan.

Dōgen, E. (2006). *Master Dōgen's Shōbōgenzō*. Translated by G. Nishijima and C. Cross. Book Surge Press.

Fish, S. (1989). *Doing what comes naturally: Change rhetoric and the practice of theory in literary and legal studies*. Oxford: Clarendon Press.

Giroux, H. (2018). *American nightmare. Facing the challenge of fascism.* City Lights Books.
Herbart, J. F. (2021). *The science of education, its general principles deduced from its aim and the aesthetic revelation of the world.* Legare Street Press.
hooks, b. (2014). *Talking back: Thinking feminist, thinking black.* Routledge.
Jaeger, W. (1986 [1944]). *Paideia: The ideals of Greek culture, vol. III. The conflict of cultural ideals in the age of Plato.* Oxford University Press.
Jaeger, W. (1965 [1939]). *Paideia: The ideals of Greek culture, vol. I. Archaic Greece: The minds of Athens.* Oxford University Press.
Kant, I. (2012). *On education.* Dover Publications.
Latour, B. (2017). *Facing Gaia: Eight lectures on the new climate regime.* Polity Press.
Mouffe, C. (2005). *On the political.* Routledge.
Næss, A. (1986). The deep ecological movement: some philosophical aspects. *Philosophical Inquiry* 8 (1/2): 10–31. doi:10.5840/philinquiry198681/22.
Orellana, P. D. and Michelsen, N. (2019). The new right: how a Frenchman born 150 years ago inspired contemporary nationalism. *New Statesman*, 3 July. Retrieved from www.newstatesman.com/politics/2019/07/the-new-right-how-a-frenchman-born-150-years-ago-inspired-contemporary-nationalism.
Rancière, J. (1991). *The ignorant schoolmaster: Five lessons in intellectual emancipation.* Stanford University Press.
Rancière, J. (2007a). *On the shores of politics.* Verso.
Rancière, J. (2007b). *The politics of aesthetics.* Continuum.
Rorty, R. (1980). *Philosophy and the mirror of nature.* Blackwell.
Säfström, C.A. (2020). *A pedagogy of equality in a time of unrest: Strategies for an ambiguous future.* Routledge.
Säfström, C.A. (2021). Conclusion: Are you awake yet? In K. B. Petersen*et al.* (eds), *Rethinking education in the light of global challenges: Scandinavian perspectives on culture, society, and the Anthropocene.* Routledge.
Säfström, C.A. (2023a). *Education for everyday life: A sophistical practice of teaching.* Springer.
Säfström, C.A. (2023b). The forgotten language of public education. In C. A. Säfström and G. Biesta (eds), *The new publicness of education: Democratic possibilities after the critique of neoliberalism.* Routledge.
Todd, S. (2023). *The touch of the present: Educational encounters, aesthetics, and the politics of the senses.* SUNY Press.

4

SCULPTURAL KNOWING

Barbara Knežević

Introduction

In the context of the current anthropogenic climate crisis and ecological collapse, there is a need to engage alternative tools for knowledge production that disrupt dualist world views. Dualist world views have encouraged human exceptionalism in the form of oppositional positions such as mind and body, nature and culture. Knowledge making that fosters empathic material engagements that recognises the life and agency of non-human others; materials and objects, plants and animals, minerals and other entities could provide a small tool with which to make out of concepts of human boundedness and anthro-exceptionalism. Contemporary sculpture has recently engaged in a renewed attention to materiality and haptic and embodied practice that could be positioned as a knowledge practice of sculptural knowing. Sculptural knowing is as an embodied practice that proposes knowing with materials as a method of artistic production that is empathic, embodied, sensate, and full of regard for the agency of materials, allowing space for an artistic imagining of worlds that are more than human. It is a way in which to challenge dualist, patriarchal colonial positions in the production of knowledge to create different considerations of human relations to other humans, animals, matter and objects. Sculptural knowing is an epistemological mode with which to make a way out of dominant dualist and linguistic models of knowledge creation.

Prevailing linguistic and semiotic models of knowledge production correlate to dualist world views that have served, in Western culture, to privilege language over matter along with other binary, hegemonic oppositional positions. Models for knowledge that privilege linguistic or semiotic modes of being in the world have been problematised and critiqued in recent feminist

DOI: 10.4324/9781003456636-5

and Posthumanist discourse, due to how these models of knowledge creation uphold political and social privileges and hegemonies that track across dualist splits in how the world is conceptualised—such as mind and body, nature and culture. Feminist and Posthumanist discourses point to the enmeshed nature of the human in matter, the human subject as part of but not exceptional within systems of ecology, environment and the non-human world. Drawing on the work of Rosi Braidotti, Lucy Lippard, and Astrida Neimanis, the inseparable condition of the matter of the human body from the matter of the world is framed as being realised and explored in the knowledge practice of sculpture.

Sculpture as a tool for dismantling human exceptionalism is framed by outlining the recent discourse in posthumanism that has critiqued dualist conceptions of the world and the linguistic turn. Sculptural knowing is proposed as an alternative, praxis-led model for making out of dualist systems. The example of making clay sculptures in the studio environment is used to demonstrate how sculptural knowing accesses ancestral and feminist methods and how materially led, collaborative knowledge practice that defies the conception of the boundedness in relation to the human body. This is viewed through the practice of sculptural clay work referencing Lucy Lippard's writing in relation to art of prehistory and Astrida Neimanis's conception of watery porous bodies she calls 'Hydrofeminism'. Sculpture, and the notion of sculptural knowing is positioned as capable of providing empathy for the non-human, material world. This is described by first examining the definition and roots of the term empathy, and then by referencing an exhibition in a gallery environment by the artist Liz Magor, whose sculptural works provide empathic, sculptural knowing with objects that are mass produced, discarded, in concert with objects and scenarios rendered by the artist.

Alternative concepts for thinking and living in the world are required and a resistance to dualist conceptions of the world, a resistance advanced by Posthumanist scholarship, are placed in the context of artistic sculptural praxis. What can haptic and materially formed knowing produced in the making and experience of sculpture offer the maker, viewer and human participant in reattuning to embodied knowledge? Sculptural practice and knowing is proposed as a pathway to promoting empathic understandings and making of knowledge with not of, the non-human world. This could be posed as a contribution to and proposal for ancestral, alternative conceptions of knowledge making and to the creation of embodied empathy relation of human subjects to non-human that is particularly vital in context of climate crisis and global ecological collapse. There is a need for recalibrating Western thought traditions in relation to the non-human world. Rather than instrumentalising haptic, sculptural knowledge practices as tools with which to engage resilience and individualistic responses to crisis, sculpture and sculptural knowing can be a device with which to create alternative ways to live and exist in the world collaboratively with matter and things, in the context of anthropogenic climate crisis.

A Return to Material

Recent posthumanist and feminist discourse has pointed to the dominance of linguistic and semiotic forms of knowledge creation in the twentieth century. The dominance of linguistic forms of knowledge, identified as 'the linguistic turn' has in the last 20 years been challenged by the 'material turn' that has found expression in Graham Harman's work on object-oriented ontology, the new materialism of Jane Bennett, and posthumanities of Rosi Braidotti and Donna Haraway. The work of these practitioners and discourses has become prominent and important in the discourse of art in the last twenty years as artists and practitioners embraced a renewed attention to materials in philosophy as agential and lively. Rosi Braidotti, responsible for the neologism 'neo materialism', describes the genealogies of posthumanist and new materialist thought and discourse in *New Materialism: Interviews and Cartographies* (Dolphijn and van der Tuin, 2012) and how this discourse has arisen in reaction to the hegemonic position acquired by Post-Structuralist linguistics of semiotics and psychoanalysis.

Braidotti describes neo materialism as a discourse that 'refuses the linguistic paradigm, stressing instead the concrete yet complex materiality of bodies immersed in social relations of power' (Dolphijn and van der Tuin, 2012, p. 21). Braidotti contends that 'As a meta-methodological innovation, the embodied and embedded brand of feminist materialist philosophy of the subject introduces a break from both universalism and dualism' (Dolphijn and van der Tuin, 2012, p. 21). This meta-methodology of embodied subject, and embedded, situated knowledge is posed by Braidotti as an alternative to dualist, linguistically led knowledge, and this methodology can be identified in the practice of sculpture. In the practice of sculpture, matter's meaning iterates itself transversally across the territories of bodies, matter, politics, thought and economies to combine to create a materialised sculptural work. Sculptural exists in a plurality, challenging the reductive hegemony of binary and dualist versions of the world. It acknowledges, as Braidotti says, the material existence of the subject and the embodied nature of being.

There is an urgent need to reconsider the relations between the human and the more than human world in the context of climate crises but also in terms of movements for human social justice. Braidotti sees these movements as intersectional:

> As a brand of vital materialism, posthuman theory contests the arrogance of anthropocentrism and the 'exceptionalism' of the Human as a transcendental category. It strikes instead an alliance with the productive and immanent force of zoe, or life in its nonhuman aspects. This requires a mutation of our shared understanding of what it means to think at all, let alone think critically.
>
> *(Braidotti, 2013, p. 66)*

The notion of thinking together and forming alliances with materials that Braidotti names in the above paragraph is shared in the practice of sculpture, where knowing is co-produced with materials and with the outcome of more than human empathy. The work of Braidotti and others in the post humanities position the divisions posed between subject/object, nature/culture, matter/human as false binaries, binaries that are actually not representative of what we experience materially in the world, but that are imposed on us in discourse that have deep political and social consequences. The political and social consequences of these dualist divisions are that they serve to uphold hierarchies that not only can be felt across material relations, but that they are replicated in human social and political relations.

The project for dislodging dualist positions and human exceptionalism in the post humanities as Rosi Braidotti notes, is not just the decentring of any human it is the decentring of the notion of Universal Man. Universal Man is positioned by Braidotti, and more broadly in the post humanities as a unified white male subject, probably European or American, who in the context and discourse of art history is in possession of unique genius and possesses the authorial qualities and privileges that elevate him above his materials and conversely above other people, by way of an access to knowledge that others cannot possess. As Braidotti notes, universal man 'is implicitly assumed to be masculine, white, urbanized, speaking a standard language, heterosexually inscribed in a reproductive unit and a full citizen of a recognized polity … How non representative can you get?' (Braidotti 2013, p. 65).

Alongside this, the epistemological reliance on language and discourse and its correlation with the mind, in the mind and body dualism, that has shaped Western thought means that other forms of knowledge, based in material, haptic and embodied engagements have been strategically devalued. This devaluation disproportionately affects the knowledge practices of those not in the hegemonic position. How can sculptural knowing as an epistemological practice present an alternative embodied form of knowledge production and acknowledge ancestral ways of engaging with and knowing the world while dismantling the devaluation of materially led and embodied practices of knowing?

Dualist forms of knowledge creation that falsely separate mind from body and nature from culture are prevalent in educational systems and in the current milieu this could also be regarded as an educational problem. Symptomatic of dualism is the reliance, even in art educational models, on language and discourse as modes that are positioned as relating to the mind and intellect. Meanwhile the body (of which the mind is inherently a part), and embodied knowledge produced haptically or materially has been deprivileged under these conditions. Manual labour, skills-based learning and practices of making are often positioned academically as crafts, trades, tools to represent ideas. Fundamentally as less intellectual—or even as less intelligent—forms of knowing the world, under these dualist conditions.

The collaborative and materially led practice of sculpture has commonalities with pedagogies that recognise the agency of students and teachers with one another and of the materials with which they work. This can be thought of as attentive pedagogy, where knowledge is acknowledged as being 'co-constituted' by all actors in a learning environment, including materials, or non-human participants. Vivienne Bozalek, Abdullah Bayat, Daniela Gachago, Siddique Motala, and Veronica Mitchell, in the collaborative publication *Socially Just Pedagogies*, write that attentiveness leads to co-constitution and becoming-with and a resistance to binary divide of the self and other in the educational setting. This attentiveness is posed by the authors as a mode that is responsible and comes into being collaboratively, with teacher, student, and materials in the learning environment. They write that 'Rather than engaging with binaries like subjects/objects, teachers/students, we explore teaching as the relational phenomenon that is constituted through intra-actions leading to a becoming-with' (Bozalek et al., 2018, p. 103).

The notion of dualism that these authors resist in their relational and becoming-with proposal, indicates how problematic the separation of knowledge in the Western tradition from the body and from matter is because of the exclusions with which it operates. Dualism has reinforced a belief and allowed human subjects in the Western tradition to consider themselves as bounded, exceptional and anterior to the non-human subject, matter, plants, animals and ecosystems. Relative wealth and privilege in Western economies has allowed for the material and bodily consequences of climate crisis to remain at a comfortable distance, for now. Meanwhile, non-Western, economically disadvantaged communities and indigenous communities have unequally and disproportionately felt the bodily consequences and bodily proximity of anthropogenic climate change.[1] Western human exceptionalism and false bodily boundedness arguably allows for a disengagement of human knowledge from empathic engagements with the human and non-human world. Flowing from the dualist position, extractionist, colonial and accelerationist attitudes towards the earth and its 'resources' leads to environmental damage, pollution of ecosystems and from there, the subjugation of the non-human world becomes distanced. The wealthy Western human subject is positioned as separate from and masterful over the non-human world as Braidotti points out. New models for existence and for thinking with and in the world are required against the backdrop of climate crisis that go beyond notions of climate resilience, and individualistic responsiveness and survival and instead engage collaborative, networked, material alternatives.

In this context, considering the need for alternative models of knowing with and in the world, how can knowledge generated from sculptural engagements—sculptural knowing—provide a praxis-led, collaborative, materially informed knowledge practice that has material and non-human agents embedded in and participating in its method and methodology? The sculptural practice of making and collaborating with materials, to produce often abstruse and uncertain outcomes has the potential to dislodge prevailing linguistic, dualist archetypes and paradigms of knowledge production. Sculptural knowing is not new knowledge, or a claim to a new knowledge practice. Rather it is a connection with older knowledge practices that stretch back into ancestral and deep time. In the following passages the knowledge practice of sculpture will be explicated in terms of how it can propose other ways of thinking and being with the world of non-human and more than human others in a way that breaks down the prevalence of anthrocentric, dualist world views.

In Praxis: Sculptural Knowing

Sculptural knowing is a form of knowledge making that occurs through the relationships and deployment of materials in collaboration with the action and intention of an artist in production of a sculptural artwork or exhibition. Sculptural knowing takes place with and for materials, objects, language and ultimately, the world. *Knowing*, is an active force, plural, unfixed and always in a state of becoming through deep, sometimes exhilarating, sometimes confounding interactions and intimacies with things. Knowing is contrasted here with knowledge. Knowing in the context of sculpture is phrased not just as an organised, institutionalised, epistemological practice complete with well furrowed hierarchies, but as a variegated affective series of relations that operate across a multitude of modalities in several contexts—potentially the studio, the university or art school, the home, an outdoor site, a gallery or museum. It can take place in the context of a group class and in communal making, or in a private studio environment, where an artist is one to one with materials and objects. Following from this, it can take place in a gallery or museum when a viewer apprehends and collaborates knowingly with a sculpture or in less formal encounters with sculptures or objects in the world. Sculptural knowing is para-institutional, in that it can take place inside an educational structure or institution but also outside it or in alternative contexts. It dwells in the borders, across them, inside the institution and simultaneously exceeds the institutions of education and art. As alternative arts educators and founders of *Dark Study* Cherry and Maloof would describe it:

> Para-institutional spaces exist besides and beyond the institution, forming alternatives while overlapping. They are peripheral and ad-hoc, part, but not part. They move beyond logics of extraction, remove barriers to accessibility, while embracing new models of knowledge transmission.
>
> *(Hermon 2021)*

Similarly, sculptural knowing may operate as a part of institutions of education and museums and galleries, but it does not depend on them for existence. Two examples of sculptural knowing are explored in this regard, one that taking place typically in the private studio with clay, and the other in the more organised and institutional context of an exhibition in a publicly funded gallery. Knowing, in sculptural knowing is framed as at least partially acquired from a non-human agent, that is the material, tools and processes and the assemblages of these sculptural works are co-produced collaboratively between the human subject and the 'things' that comprise the sculpture in a less didactic form of knowledge production.

In Praxis: Making with Bodies of Clay

Sculptural knowing is a small tool, what Donna Haraway calls a sighting device, it is 'Not a systematic overview, it is a little siting device in a long line of such craft tools. Such sighting devices have been known to reposition worlds for their devotees-and for their opponents' (Haraway, 1992, p. 295). It is a tool with which to begin to unwind the hegemonic power of dualist and linguistic systems of knowledge. In the context of sculptural knowing, the practice of making with clay in the studio is instructive as an example of how this knowledge is constituted. Knowing, in the context of making sculptures with clay is a making-with materials that has the potential to awaken or re-instate latent, fundamental material shared sensibilities and constituent parts between maker, material, form and viewer of a sculpture. This process and engagement with clay as matter in sculptural practice does not necessarily impart new knowledge but awakens something fundamental—the shared material constitution of both the body of the material and the body of the artist and the long history of human engagement with this material for making sculptures, objects and vessels.

In *Overlay* Lucy Lippard writes about the relationship of contemporary feminist sculpture and art of the late 1970s and 1980s to art of prehistory. The term, Prehistory is worth problematising here, as it refers to human time before the written record. It is another example of how language and knowledge is correlated to the written word, and it deprivileges knowledge practices that exist as oral, or material practices. Lippard identifies the problematic conflation of women's bodies with nature and the earth, and the difficulty for feminist artists and risking engaging in essentialism in relation to the gendered body, matter and nature. Lippard reinforces the problematics of dualist world views and the social implications of these hierarchies in relation to gender. She writes that 'there is no question that the identification of woman with nature and man with culture has played a damaging role in the relegation of women to inferiority' (Lippard, 1995, p. 41). Lippard describes the potential of engaging with and revisiting ancient sculptural practices as a way of dislodging the prevailing knowledge systems that relegate matter to subjugation—along with women, indigenous people, queer people and anyone outside the hegemonic power as Braidotti outlines. The power of revisiting ancestral sculptural knowledges according to Lippard is part of rehabilitating art history and human history and acknowledging the unboundedness of the human in relation to material.

Making sculpture with clay activates an ancestral practice that pre-dates the imposed divisions of mind and body, nature and culture. Lippard quotes Gina Blumenfeld writing that the false divisions between nature and culture that have been pursued in Western culture have resulted in patriarchy, and the process of depositing 'unwanted and unruly' qualities of nature with the subject of the woman. Blumenfeld writes:

> Today we stand at the end of this long process of internal and external subjugation … the antagonistic split between man and external nature as well as that between humanity and its own internal nature (which includes the polarization between woman and man) are of a whole cloth and must come together.
>
> *(Quoted by Lippard, 1995)*

Clay is often referred to as the 'body of clay' an anthropomorphising term that relates to both its plasticity and to the historical relationship of clay to the formation of vessels. Clay has a relationship to the relic, the artefact and human history, and it has the qualities of plasticity, anthropomorphic qualities, of body, malleability, softness, transformation, and vitrification. Clay has the qualities of a geological process, of intense heat, of melting, of transformation of matter into a different molecular state, into a state of change. Clay will only tolerate certain temperatures and conditions before it will break apart and transform into something other than the intended form. Once fired, ceramics are an enduring and resilient material often the remaining material evidence on archaeological sites. The process of making with clay gives the maker proximity to and embodied experience of earth, of a proximate embodied geology, but it also connects the maker to the history of humans and of human sculptural practices that extend into prehistory. This can be characterised as a latent knowing, working with clay is a process that approximates geological and deeply earthly systems while also connecting with the narratives of human history, politics, cultures, and material practices. This could be thought of metaphorically as clay memory.

Clay memory also describes the molecular qualities of clay. It has a material memory that means that it can often bend back on itself on firing, into shapes in which it was previously formed. It is as though the clay permanently records the actions and intentions for it that were tried and scrapped, deep within its molecular structure, bending and twisting itself back into deleted forms when it is fired. The artist Phoebe Collings-James writes evocatively about this memory in conversation with Julia Phillips and Dr Jareh Das. She describes making with clay as a collaboration with materials, a conversation, a reciprocal connection, she notices that clay feels and responds to its new form. 'it's reciprocal … Reciprocity also appears through sculpting and shaping: the clay then decides how it feels about its new form, potentially shifting, cracking or remembering the other configurations that it had before'

(Collings-James et al., 2021). Phoebe Collings-James describes a process and experience with clay that is an exchange where dualist divisions come away, acknowledging the agential nature of the clay in relation to previous forms and iterations. It is not a passive material. Working sculptures in clay in this sense embodies sculptural knowing; unfixed, in progress, transformation, in negotiation and constantly in formation.

My practice-based experience of the embodied process of creating a ceramic vessel or form means that I am often conscious of co-making and collaboration with clay when I use this material. I know sculpturally and haptically that it is a conversation, a dialogue between my intentions, my physical body, the past, history and the clay body itself. My hand meets the wet clay, a smooth, creamy and yet grainy texture, processed and presumably mined from an earthly source and extruded into industrial blocks of pre-sieved and mass-produced quantities, sold in a ceramic supplies store. I cut a chunk of clay from the oblong-shaped bag, using a wire cutter that passes with damp resistance through the quiet mass to form a square piece. I hold the piece in my hand and move it around before tamping down the edges on a wooden board to form a circular piece. The clay stains and covers my hands, I can't see it, but it is merging into my pores, under the fingernails and the particles and minerals enter my body through the dermis. My hands and arms know how far to stretch and form the plastic clay body, to move with it, to push and pull it but only within the boundaries of what the material itself will allow and will construct with me. Clay instructs me, and I instruct it, neither of these positions exceeds the other. Together we make a form, and sometimes, this form is fired.

This porosity of the body in relation to working clay in sculpture recalls the recent work of Astrida Neimanis, and her proposition of a hydrofeminism, that invokes water, and human bodies as a model for thinking the human subject and body as unbounded in relation to the world, ecosystem and non-human entities. Neimanis writes about the cyclical embeddedness of human bodies in ecology, as a way of challenging dualist, patriarchal and Western positions and the false divide between humans and nature. Neimanis does this particularly in relation to water. Neimanis writes in *Hydrofeminism*:

> We are all bodies of water. To think embodiment as watery belies the understanding of bodies that we have inherited from the dominant Western metaphysical tradition. As watery, we experience ourselves less as isolated entities, and more as oceanic eddies …
>
> *(Neimanis, 2016, p. 27)*

Neimanis's trope of water corresponds to the experience of sculptural knowing with clay. With clay the body becomes unbounded, porous and in close touch with matter, it can be felt as part of the body. Water is an example of how the philosophical boundedness of the human body can fall away when considered in relation to a substance that makes up as much as 60% of the human body. This too happens when working with clay. Neimanis instrumentalises the trope of water in her work as a tool for thinking out of and beyond the Western tradition of considering human bodies as separate from nature. Human bodies are water, are mineral, are geological, are vegetal, we are animal. In the context of sculptural knowing with clay, this unboundedness means that instead of asking ourselves to have empathy for the 'non-human' we are instead asking for empathy for ourselves, our own bodies and matter. 'Ourselves' are radically refigured as unbounded bodies fundamentally part of and made of more-than human worlds. Through the practice of sculptural knowing with clay, an enmeshed embodied being and understanding of the relationship to the material and non-human world may be reached.

In Praxis: Sculptural Empathy

Sculptural knowing and practice produces knowledge through the body engaged with materials. This haptic and materially led knowing has the potential to foster empathy for the non-human world. This empathy for the non-human and matter of the world presents a break with dualist thinking and separations and dislodges entrenched thinking about human exceptionalism in relation to the material world. Empathy in this case is produced in sculptural knowing not by asking for empathy for things that are anterior to the human, but by encouraging a cosmology that sees the human subject as deeply and primarily material. Aesthetic empathy was explored in the work of Vernon Lee in the early twentieth century, and was characterised by Lee in terms of a discussion of a human experience of apprehending a mountain. Lee (pseudonym of Violet Paget) writes in *The Beautiful*:

> And it is this complex mental process, by which we (all unsuspectingly) invest that inert mountain, that bodiless shape, with the stored up and averaged and essential modes of our activity—it is this process whereby we make the mountain raise itself, which constitutes what, accepting Prof. Titchener's translation[*] of the German word Einfühlung, I have called Empathy. [*] From ἐν and πάσχω, ἔπαθον. The German word *Einfühlung* 'feeling into'—derived from a *verb to feel oneself into something.*
>
> *(Lee, 1913, p. 61)*

The sense that empathy is a *feeling oneself into something*, and the way that Lee describes this in terms the possibility of feeling into the enormity and deep time geology of a mountain corresponds to the earlier passages concerning my own experiences of making sculpture, a feeling into clay. This *feeling into* can be placed in context of the experience of encountering and entering into any material or forms of sculpture. Lee writes further that empathy relates to the ego, where paradoxically the projection of the subject's ego into the object is a momentary 'abeyance' of the ego, where a merging of the qualities of the subject and object takes place (Lee 1913, p. 61). The proposition made by Lee is that even when a human subject is engaged in apprehending a mountain, it is possible to lose the 'ego' or subject-hood of the self. The self becomes subsumed into the object being apprehended. In the context of sculptural knowing this could be considered to signal a moment of recognition of the shared materiality of the human body and the mountain, that erodes the sense of human exceptionalism, an empathic materially led encounter taking place on a deeply material level. In context of sculptural knowing, the feeling into is not a co-incidence, or a chance encounter. The feeling into empathy with a sculpture is an encounter that is entered into by a viewer, and deliberately orchestrated by the artist.

The work of sculptor Liz Magor embodies the empathic potential of contemporary forms of sculpture and sculptural knowing as a practice. Her work takes place against the backdrop of the contemporary milieu, where the human relationship to material and objects has become one that is awash with one use items, plastics, packaging and of a general brevity in the everyday encounter with things that are often characterised as one-use items. Martin Herbert writes on contemporary sculpture that 'Right now, widespread evidence suggests that what they want to make relates to the tides of materiality around us sifting landfills, in a way, or sifting archives and to a parallel drift, towards life online, whose weightlessness deemphasises materiality' (Herbert, MacDonald, and Strang, 2013, p. 46). Herbert points to the move towards the strong desire to make sculptural artworks that are assemblage-like and take on the appearance and energy of the tides of materials that wash around the contemporary human subject. The sculptural assemblages evident in the practice of Liz Magor consider the value and treatment, the affect and emotive call of vibrant, transversally active materials and objects of various cultural value. In so doing, Magor's assemblages bring a *feeling into* the things that are often considered valueless.

Magor's work was the subject of a solo exhibition in Dublin at the Douglas Hyde Gallery in July 2023. The exhibition was titled *The Rise and Fall* and was comprised of a large array of clear mylar boxes placed on the gallery floor, containing and holding disparate items such as rat skins, holographic paper, the insides of plush toys, patterned fabric. Suspended from the ceiling of the gallery were two long hanging sculptures, one featuring a tan-brown-coloured silicone cast of a stuffed monkey and draped fabrics and other materials. The other features a hanging dark brown coloured silicone bear-like stuffed animal, garlanded with taffeta and tangled threads, a grey wool fabric drapes down onto the floor. There was a pale dusty grey jesmonite cast of a cardboard box resting against one wall, with a delicate taxidermal bird lying on its side on top titled *Wasted.*

On another wall was a distressed grey wool hooded coat, a thrifted item, with seashells, hair, confectionary wrappers, fibres and other hard to discern objects spilling from one pocket. Scattered across this coat is evidence of careful embellishment of what look like moth holes alongside streaks of silver pigment or paint. One sleeve held what looks like a makeshift flag. This work is titled *Perennial.* In the sunken area of the gallery rested a light blue silicone impression of a large stuffed lion, nestled in a fleece blanket. The blue lion had a mane made of white nylon fake hair and one visible eye socket seems to be burned out, emptied of its eyeball and softly singed around the edges. The lion rested on a large, low display platform in muted dusty pastel tones that appears to be a bed-like divan structure, complete with pleated fabric skirting that either resembles or is a material used in domestic interiors. This structure recalled the reclining or slumbering human body, though the lion was accompanied here in this scene by other entities. These other entities are various in terms of their materials and their provenance, listed as painted plywood, fabric skirting, silicone rubber, artificial hair, acrylic throw, woollen blankets, silver fabric, linen, jewellery boxes, costume jewellery, packaging materials. This work is titled *Coiffed* (2020).

The objects in *The Rise and Fall* were individually familiar, but together they formed a strange and emotive tableau that evoked something domestic and intimate. Magor dysregulates the fluency of familiar, domestic readings by rendering certain objects in incongruent materials. The stuffed animals, 'stuffies' as Magor refers to them, are not soft plush toys, but are all facsimiles[2] cast from toys then rendered in silicone. The traces of the clumped, wet fake fur as it congealed under the wetness and weight of the casting material are recorded in the silicone impression and this is what gives away this object's oddity. The array of things invites a viewer's attentiveness, placed as they are on display plinths of various sorts, though they produce their strangeness and dissonance in concert. In her collected writings and interviews *Subject to Change*, published in 2022, Liz Magor writes that she considers the position and status of the objects she uses from the world, that these objects operate by instigating affect in humans, both positive and negative. Says Magor 'In the studio I might rearrange the relationship between things in order to increase their power, or I make adjustments to restore their depleted importance. I always assume that material is co-operative, and process is the way to reach and understand the latent intelligence of things' (Magor, 2022, p. 290).

Magor indicates her consideration that the non-human world holds a latent intelligence, that is not animalism, but some other kind of knowing that acknowledges a co-operation between maker and material. This knowing is revealed or highlighted by her sculptural practice, and an engagement in materials revealing something hidden or masked about objects. Perhaps it is a revelation of their autonomous state, unburdened of human desires, and their potent ability to enact their own affinities when unleashed in a sculptural conversation between artist and viewer. Magor refers to the transformative and revealing process of the sculptural process and gestures to what engaging with objects and materials in the process of sculptural making can allow the artist and the viewer to know about the nature of these things—their 'latent intelligence' (Magor, 2022, p. 290). Emily LaBarge writes in Artforum that she considers this a process not of revealing the artistic or other human desires but of a stripping away of these demands, allowing them to act out their own agencies. LaBarge writes on Magor's work that 'Even the most forsaken things do not disappear when our backs are turned—a reality both ecological and ideological. That includes art objects and the humble material resurrections they offer, if we pay attention. How ordinary, how remarkable, how enduring' (LaBarge, 2023).

In Magor's work, it could be said that making and act of transformation is not about imposing some artistic logic or wish upon the things being 'transformed' but indicates a situation where the artist is engaged in sculptural process to reveal, to discover, to enhance what is already there. This working-with, making-with ideology that is less anthropocentric, and less dualistic in its approach is indicative of sculptural knowing, a collaboration, a re-awakening of a human attention to what is already there and the human existence as part of the material world is revealed by scrutiny—attentiveness—and care. The properties of the materials are unveiled by an intensive and attentive engagement with them, that is at once transformative and illuminating, and acts in concert with the materials and objects, but is not intent on imposing or radical alteration to objects or materials.

The consequences and the potential of empathic sculptural knowing, of the sort that is revealed in the contemporary sculptural artwork such as Liz Magor's resides in its activation and acknowledgement of the collaboration and shared material fealties of human, material, animal and extra-human engagements that take place in the context of sculpture. The implications of this knowledge and acknowledgement of material vitalities are a renewed empathy and consideration of the more than human entities that comprise sculpture. This has the potential to spread outward into engagements and relations with non-human entities in broader contexts. Sculptural knowing can be proposed in this sense as a method with which to challenge dualist, patriarchal and colonial positions in the production of knowledge to create different considerations of human relations to other humans, animals, matter and objects.

Sculptural knowing can potentially engender empathy with things that are more than human in contexts outside artmaking and the exhibition of art. Empathic responses and the awakening buried material and fealties in the context of the now feels both vital and urgent.

Imagining the notion of responsibility in relation to materials, in the contemporary context—where every material engagement has a consequence, and each human is partly responsible for this animate relation—has implications for how knowing might operate differently in context of climate crises and late capitalism. These implications are brought to bear in the work of Liz Magor. A sense of responsibility to know and care differently even for the discarded mass-produced things that wash around the contemporary human subject in huge endless tides is a consequence of the knowing produced by and revealed by her sculptures.

Conclusion

The empathy produced through sculptural knowing has the potential to engage broader empathy for non-human life, ecosystems and materials, by recognising the shared human materiality and enmeshed nature, the unboundedness of the human body in relation to the world. The sculptural knowledge system questions human exceptionalism by reminding the human subject that it is materially constituted and porous in the world. This is offered up not as new mode of knowledge but an ancestral one. Sculptural knowing as a practice provides a glimpse both into the past practices of living in a porous and co-constituted condition with the non-human world, and into a potential future, where it will no longer be possible to deny the enmeshed and porous human relationship with the ecological systems that are in crisis as climate crisis comes to affect all human bodies. Sculptural knowing could be proposed as a break with dominant, academic paradigms of knowing the world, a break with dualism, a break with versions of rationalism or idealism that have served to promote the idea that human beings exist exceptionally and apart from other entities. This break is needed, instead of the impulse to resilience and adaptation as a mode, new/old ways of thinking are required to avoid the oppositional, dualist separations that have perpetuated a lack of fealty with non-human and materials.

Sculptural knowing is a haptic, process-led, and embodied pursuit that can reveal a deep fealty and shared relation with materials, and consequently an empathy with the non-human world. In the context of ecological crisis, and still prevailing anthropocentric attitudes towards materials as tools, resources, or consumables for human exploitation, sculptural knowing has the potential to awaken the latent knowing—to reconnect or rehabilitate in the individual and collective Western human subject the relationships to materials and non-human objects and matter. These are minor narratives, siting devices, with sculpture and materials, examining small and humble acts of haptic joy and embodied material knowing that may open up potentials for ways of understanding the material world with a decentred human knowing and acknowledgment of the collaborative and deeply networked nature of making with materials. These are small moments of acknowledgments of the vitality of materials that can take place in the production and dissemination of sculptural work.

Empathy can be produced both in the act of making and experiencing sculpture. In both instances, knowledges are co-produced, collaboratively with human and non-human entities. This process of co-production of knowing can engage an empathy that is embodied but does not just call on the anthropocentric or allow an anthropocentric empathy. The nature and structure of the empathy produced by way of sculptural knowing with objects and materials allows for empathy that respects and acknowledges difference while engaging care and appreciation for the matter and objects. This empathy locates and highlights the deeply material fealty between the human body, largely watery (Neimanis, 2016, p. 27) and composed of minerals, and the environment, the matter, the things that compose and surround the human body. Human bodies are not bounded separate entities from the materials and things of the world and environment, human bodies are *from* and *with* matter, porous and membrane like in constant and perpetual exchange with the chemical and material world. Sculptural knowing enhances and awakens in the maker and the viewer this empathic state of material bodily kinship in the studio, the University, the gallery and in other non-institutional contexts. It does this through the recall of material touch, through the embodied spatial encounter, and through this a close observation and scrutiny, a *feeling into.* It is this blending, where bodies as viewers are rendered back as equally material and mineral as the things we see and circulate amongst. The constructs and boundaries that prevent us from accessing the empathic with things that are more than human can be eroded in this space of the sculptural artwork, through the production of sculptural knowing.

Notes

1 Astrida Neimanis (2016) describes the uneven disproportionate effects of pollution, climate crisis, and ecological collapse on indigenous people through the lens of human breastmilk in Innuit communities, whose health is impacted by the aggregation of Western produced toxins and pollution in the Arctic. 'Consider that due to cold temperatures and little sunlight, persistent organic pollutants (POPs) flowing from the industrial and agricultural wastes of far-flung rich, Westernized outposts break down slowly in the Arctic. A thumb-sized piece of *maktaaq*, a staple in the Inuit diet, contains more than the maximum recommended intake of PCBs for an entire week. As a result, Innu women's breast milk is an especially toxic substance, absorbing the liquid runoff of a global political economy that produces vastly divergent body burdens. The inequalities of neo-colonialist globalization course through waterways at scales both individual and oceanic. Nursing one's young becomes a complex congeries of questions in which we all are implicated, rather than an issue for the biologically essentialized, lactating woman alone. The flows of global power meet the flows of biomatter.'

2 In an interview with Liz Magor conducted on 11 July 2023 in the Douglas Hyde Gallery, before the opening of 'The Rise and Fall', Magor stated that 'I make everything, but I have a really loose attitude towards how close it gets to the real.'

References

Bozalek, V., R. Braidotti, T. Shefer and M. Zembylas (eds). 2018. *Socially Just Pedagogies: Posthumanist, Feminist and Materialist Perspectives in Higher Education.* Kindle edition. Bloomsbury Academic.

Braidotti, R. 2013. *The Posthuman.* Polity.

Collings-James, P., J. Phillips, and J. Das. 2021. How Clay Is Connected to Our Bodies. *Frieze* (223). www.frieze.com/article/how-clay-connected-our-bodies.

Dolphijn, R. and I. van der Tuin. 2012. *New Materialism: Interviews & Cartographies.* Open Humanities Press.

Haraway, D. J. 1992. *Promises of Monsters: A Regenerative Politics for Inappropriate/d Others.* Routledge.
Herbert, M., F. MacDonald, and M. Strang. 2013. *Thinking is Making: Presence and Absence in Contemporary Sculpture: the Mark Tanner Sculpture Award.* Black Dog Publishing.
Hermon, R. 2021. Enmeshed in the Borders: The Para-Institutional Practices of Alternative Arts Education Online. *A Peer-Reviewed Journal About* 10 (1): 57–69.
LaBarge, E. 2023. Liz Magor Focal Point Gallery. www.artforum.com/events/liz-magor-3-251369/ (accessed 14 June 2023).
Lee, V. 1913. *The Beautiful: An Introduction to Psychological Aesthetics.* Cambridge University Press.
Lippard, L. R. 1995. *Overlay: Contemporary Art and the Art of Prehistory.* New Press.
Magor, L. 2022. Subject to Change: Writings and Interviews. *Text/Context: Writings by Canadian Artists.* Concordia University Press.
Neimanis, A. 2016. *Bodies of Water: Posthuman Feminist Phenomenology.* Bloomsbury Academic.

5

ARCHIPELAGO AS FORM

Education in a Post-climate World

Glenn Loughran

Introduction: What Kind of Event Is Climate Change?

Events are important aspects of our lives. Sometimes, they are easy to define and anticipate and sometimes hard to see. Regardless of how they are perceived, they are by nature significant; they register as significant in our experience and memory of them (McKenzie and Porter, 2016). Historically, events are often identified by a genuine and representable sense of before and after, for example, the storming of the Bastille or, more recently, the global impact of the COVID-19 pandemic. The theorisation of events has a long tradition in philosophical discourse, going back as far as Plato and connecting to modernity through Hegel, Deleuze, and Badiou. Many of these philosophers and theorists analyse historical events and our responses to them to delineate an evental form or an evental structure. Paradoxically, within these analyses, the event is often defined against structure, as anti-structure, and yet events are also a phenomenon that we, as humans, are hardwired towards. This is because events usually signify change, movement and novelty. As pointed out by William Sewell, 'Events may be defined as that relatively rare subclass of happenings that significantly transforms structures. An eventful conception of temporality, therefore, is one that takes into account the transformation of structures by events' (Sewell, 2005, p. 100).

Given these characteristics, it is strange that the event does not have much traction within contemporary theorisation of the Anthropocene, a significant event that has changed our perspective of the world. There are several possible reasons for this, but one that is most appropriate for this publication concerns the nature of the event of climate change itself, and specifically, how it breaks with the traditional structure of the event as cause and effect, linear, and universal (Sewell, 2005). In a key text on this subject, Tavory and

DOI: 10.4324/9781003456636-6

Wagner-Pacifici (2022) argue that because of its 'oscillating temporalities' and scales, climate change obfuscates normative definitions of the event, unfolding across 'years, sometimes decades, sometimes centuries' (p. 1). Crucially, these temporal leaps have contributed to the 'ambiguous nature' of the event of climate change, an ambiguity that leaves an open door to climate denial, and the 'triumphant relegation of climate change to normative, non-threatening reality of everyday life' (Tavory and Wagner-Pacifici, 2022, p. 6):

> Climate change is a strange event. Marked more by a slow tearing than a sudden rupture, it is thrown into the future even as its pasts and presents make themselves known through extreme weather events. As an averaging of temperatures and weather patterns, it is hard to clearly pinpoint a specific moment, providing an opening for climate denial.
>
> *(Tavory and Wagner-Pacifici, 2022, p. 9)*

At stake in this analysis is an attempt to identify a specific set of research problems related to the ambiguity of the event of climate change, exploring how different disciplines and movements tackle climate change's 'ambiguous rupture', asking, 'If climate change is an ambiguous event, how do different groups conceptualise it? Where do they locate rupture? And for whom?' (Tavory and Wagner-Pacifici, 2022, p. 2). Describing the event as a relationship between 'rupture and form' the authors analyse how distinct disciplinary fields have cultivated eventfulness as form, in response to climate change. Within all three modalities are forms of action, reflection, and communication that tie climate change's past and present experiences to its speculative futures. They are: (1) *scientific eventfulness*, which aims to unpack how scientific language simultaneously asserts and hedges the impact of climate change; (2) *radical eventfulness* which aims to develop a linguistic performativity around the urgency of climate change, understood most clearly through environmental activism, and finally (3) *sensible eventfulness* which aims to normalise the impact of climate change, through its relation to the global economy, what they call in the end, a 'non-event' (Tavory and Wagner-Pacifici, 2022, p. 20).

Analysing a selection of scientific texts on climate change, the authors identify an 'intertwining of urgency and careful hedging about the eventfulness' (Tavory and Wagner-Pacifici, 2022, p. 4) of climate change. In *scientific eventfulness*, climate change is identified as a verifiable event; it is here in the present, and yet, because of the very nature of scientific modelling, its future is uncertain. Alternatively, the *radical eventfulness* defined by environmental movements and activist groups such as *Extinction Rebellion* tends to dramatise the urgency of climate change as an event that has happened, with a very predictable and existentially dark future. Finally, *sensible eventfulness* emphasises the normative 'rational and responsible' approach to climate change, characterised by policy changes such as the Green New Deal. While the impact of such policies cannot be disputed, they are problematised by the authors as a non-event: 'If rupture is constitutive of an event, then it cannot be sensible—it requires new modes of action, new footings, an urgent mood. If it is sensible, and "responsible" then it cannot be a fully-fledged event' (Tavory and Wagner-Pacifici, 2022, p. 9). Although there are crucial differences between these modalities, what the first two have in common that the third modality, *sensible eventfulness*, does not, is their commitment to directly communicating the event of climate change. Notably, within the post-climate aims and ambitions of this publication, both modalities assert that 'the climate crisis *is* happening, and *has been* happening' (Tavory and Wagner-Pacifici, 2022, p. 6):

> Time seems to work differently in the midst of a crisis. Often, the metaphor we defer to is the image of a clock. We are running out of time. We are sounding the alarm. We only have ten years left to save the world. The climate crisis has upended our understanding of time, and reframed all human activity as a fight against the clock. In the western imaginary, the climate crisis is a bomb that is waiting to go off. In truth, the situation is much, much worse. The bomb has already detonated. Or, rather, there are millions of bombs and they are detonating all of the time. People are dying now. People are being displaced now
>
> *(Tavory and Wagner-Pacifici, 2022, p. 6)*

When it comes to the specific nature of art and education in our post-climate condition, it is helpful to consider types of evental form that might be appropriate to address the event of climate change. When speaking about educational form we are concerned with how certain elements; space, time, resources, content, are organised and composed in such a way as to support the exploration of knowledge and truth in our world. Initially, the clearest connection to art and education in this analysis would be through the first two modalities, *scientific eventfulness* and *radical eventfulness* as the both confirm the event of climate through direct action and critical analysis. However, if we were to decouple the concept of the sensible from the 'rational and responsible' and replace it with the aesthetic–political meaning ascribed to the sensible by Jacques Rancière, we could recuperate an understanding of educational form as *sensible eventfulness* that 'revolves around what is seen and what can be said about it, around who has the ability to see and the talent to speak, around the properties of spaces and the possibilities of time' (Rancière, 2013, p. 13). Similarly, reflecting on the nature of educational form in the work of Walter Benjamin, Tyson Lewis writes:

> I use the term form to mean any sensuous or non-sensuous shape or configuration that has the potential to make something knowable, visible, audible, recognisable or legible. As such, educational forms point toward an ability (potentiality) with something that would otherwise be unknowable, invisible, silent and so forth.
>
> *(Lewis, 2020, p. 18)*

If the common experience of evental rupture is unknowing, then the concept of educational form can provide a space for re-knowing, through re-forming and re-shaping. In the visual arts, the idea of education as a form of sculpture has its roots in the work of German artist Joseph Beuys, who developed a concept of social sculpture around the ambition to create a 'social organism as a work of art' (Beuys, 2006 [1973]). Aiming to intervene in social situations to shape those situations towards more constructive and democratic goals, Beuys developed large-scale durational projects that addressed political and environmental issues through direct action. For example, in 1971 he developed the Organization for Direct Democracy through Plebiscite in 1971, which aimed to adopt the political form of the party as a public artwork. In 1982 he developed a five-year plan to plant seven thousand oaks in Kassel Germany. Importantly for the next generation of young German artists, Beuys incorporated educational gestures and actions into the concept of social sculpture, through performative lectures and conferences. These pedagogical events enabled Beuys to challenge traditional distinctions between art and education that too often positioned education as an inferior, didactic activity to art-making. For Beuys, education was an act of creation, an event that would produce critical discussion. In her short reflection, *Notes on the Pedagogical Turn in Art*, Kristina Lee Podevsa highlights the importance of Beuys educational work when she writes that Beuys's 'performance-lectures were not temporally bounded in a string of discrete, finite events, but were intended to prompt further discussions carried out by the audience (post-performance) in a situation' (Lee Podevsa, 2007, p. 1). In this sense then, there is a performative cross over in art and education, between an evental form and an iterative form, that lends itself to the shaping of ideas, places and subjects.

The concept of the event has been a consistent theme in education over the past two decades. Often understood as an encounter with the 'other', key thinkers in the field argue that education is inherently characterised by qualities that are unpredictable, interruptive, and subjective (Biesta, 2021; Atkinson, 2007; Britzman, 2000). For UK theorist Denis Atkinson a pedagogy of the event disrupts the status quo and works against educational norms in support of 'local ontological and epistemological states where learning can be conceived as a movement against previous modes of understanding into new modes' (Atkinson, 2007, p. 55). Informed by the work of Alain Badiou, Atkinson argues for a pedagogy of the event that opens up spaces for 'that-which-is-not yet', for encounters with real learning against representation 'that challenges the learner out of a complacency, a comfort zone' (Atkinson, 2007, p. 54). Similarly, for Gert Biesta a pedagogy of the event is premised on the need to cultivate and sustain the experience of risk in educational situations, because 'without the risk, education itself disappears and social reproduction, insertion into existing orders of being, doing, and thinking, takes over' (Biesta, 2021, p. 140). Crucially, within these educational contexts, the teacher must take responsibility for the student's exposure to risk, to be sensitive to what Debora Britzman calls 'difficult knowledge'.

Britzman's psycho-pedagogical theory of 'difficult knowledge' engages with the experience of traumatic knowledge in the classroom. Traumatic knowledge is understood as challenging content or images of violent events that may shock the student, such as the holocaust or social and political persecution. At stake in the theory of 'difficult knowledge' is an ethics and practice of negotiation around how to manage complex content, to encounter that which is other to ourselves, because 'to study the difficulty of others is actually to study how one comes to relate to the conditions of difficulty expressed, as opposed to somehow attempt to reacquire the felt experience of the other' (Britzman, 2000, p. 38). The theory of 'difficult knowledge' is helpful for thinking about how we might understand student experiences of climate change. Many of our students experience climate anxiety and stress in the face of increasingly worrisome scientific data, live reporting of specular and catastrophic weather events, and maybe more subtly, the scale and ambiguity of the event of climate change as outlined by Tavory and Wagner-Pacifici (2022). Catastrophic climate events must not be disavowed through the curriculum structure; no matter how challenging the content, the reality of the situation compels us to find ways to navigate such 'difficult knowledge' with and for our students. Crucially, we need to find ways to balance the spectacular event of climate change and the ambiguous event of climate change; as Ursula Heise has stated 'climate change poses a challenge for narrative and lyrical forms that have conventionally focused above all on individuals, families, or nations, since it requires the articulation of connections between events at vastly different scales' (Heise, 2008, p. 205).

One way to address these challenges in the classroom is by incorporating event-based methods into the curriculum, as Elisabeth Strayer has done. Similar to the work of Tavory and Wagner-Pacifici, Strayer identifies the vast temporal and global scales of climate change to be an educational challenge, alongside the spectacular and traumatic impact of historical ecological events. Providing further nuance on this she points to the concept of 'slow violence' developed by Rob Nixon. For Nixon, slow violence complicates the presentation and representation of the event of climate change as it unfolds across multiple scales, by asking 'how can we convert into image and narrative the disasters that are slow moving, and long in the making?' (Nixon, 2011, p. 3). Developing a methodological approach to these issues, Strayer turns to the concept of the event as 'a series of connected sensational moments' (Strayer, 2019, p. 22) that can address the diffuse and ambiguous nature of climate change, because 'what every event must share is an effect that travels outwards and that transcends the immediate temporality of the original occurrence' (Strayer, 2019, p. 21). To further illustrate this approach Strayer analyses the environmental catastrophe of the 1883 volcanic eruption on the island of Krakatoa in the Sunda Strait, arguing that 'world-shaping ecological events' can be used methodologically to connect 'scales that deemphasise human time and human space' and help locate our 'place within an expansive, intricately networked ecosystem' (Strayer, 2019, pp. 13–34). Within the context of such multi-scalability, she makes an important connection between events in their singularity and the plurality of creative expressions they can engender:

> Event-based thinking can grant us room to study how a single occurrence gets folded up into a number of discourses and how aesthetic forms beyond orature and literature circulate and register events differently while, nevertheless, allowing us to return to the initial occurrence. By integrating such mediums as painting, sculpture, and film into the archive of an event, we can further comprehend its lasting impact, and even its ongoingness.
>
> *(Strayer, 2019, p. 14)*

Importantly for this chapter, collage and assemblage are formal composition tools in an event-based methodology because both support 'radical juxtaposition, the scholar's act of paratactic cutting and pasting' (Friedman, 2015, p. 77). Following this, Stayer argues for the significance of peripheral, patchwork sites of engagement with the event of climate change, such as islands and archipelagos. Due to their heightened sense of precarity and vulnerability, islands are often seen as micro-representations of global ecological instability. As a result of this archipelagic thinking is central to any political, pedagogical or artistic engagement with climate change today. Archipelagic thinking can be understood as a post-colonial, theoretical discourse that emerged out of island contexts in the late eighties through the work of French Caribbean thinkers such as Martinique poet and philosopher Édouard Glissant (1928–2011). Arguing against traditional colonial representations of centre and periphery, Glissant emphasised a global relationality that posited flow and dialogue against stasis and dialectical thought'. Through this position he challenged the image of the island as an isolated, inward-looking entity, left behind by modernity (Glissant, 1997, p. 23). As an antidote to modernity's 'continental' presumptions of 'origin, unity and security', Glissant privileged the interdependent, rootless subjectivity of the archipelago, its multi-scalar processes and relational ontology:

> [C]ontinental thought ... makes us think that we see the world as a bloc, taken wholesale, all-at-once, as a sort of imposing synthesis, just as we can see, through the window of an aeroplane, the configurations of landscapes or mountainous surfaces. With archipelagic thought, we know the rivers' rocks, without a doubt even the smallest ones.
>
> *(Glissant, 2009, p. 45)*

What is important about this proposal for environmental education today, is the extent to which it enables a profound shift away from normative understandings of globalisation as standardised network towards a spatiality that connects locality and world. Taking the archipelago as a geographical form of assemblage, and assemblage as a way to think about events and event-based methods of study, the next section will reflect briefly on a pedagogical art project developed in the west of Ireland in response to climate change. This project began as a series of performative events across multiple island spaces, with multiple island communities to explore the potential of an archipelagic master's programme in art and environment in response to the event of climate change.

MA in Art and Environment

Motivated by a constellation of theoretical considerations and multi-regional contexts, the initial research stage of the project took on the form of an educational question, *What Is an Island?*. The *What is an Island?* project was initiated within the geographic context of West Cork, the political context of Brexit, and the environmental context of the Anthropocene. It began as an open-ended, experimental journey into the changing nature of islands in contemporary life and an exploration of the characteristics of artistic research and its education. The project bridged two educational programmes through this process: the BA in visual art on Sherkin Island (TU Dublin) and the MA in art and environment (TU Dublin) in the West Cork archipelago. Connecting these educational points, the project opened up an inquiry into the significance of place-based art education operating at the edges of the university. This art education links the island, archipelago and world. The form of this enquiry was multi-modal, combining live events, performative lectures and community dialogue within island communities. These events reflected on (i) the methodological potential of archipelagic thinking for art education in climate change, (ii) the development of event-based, artistic research processes, and (iii) the impact of visual arts education on isolated island communities (Loughran, 2020).

Comprising three iterations in total, the first event took place in 2018 across the islands of Sherkin, Heir and Long. It explored the value of archipelagic thinking for art education at a time of geo-political tension in the Brexit negotiations. Central to the first iteration was an interrogation of the myth of the island as an isolated isle in a vast ocean, and an affirmation of the relational dimension of the archipelago. Responding to the question *What is an island?* Three local artists were commissioned to develop work on one of the three islands, they were: Sherkin Island (artist Mona O'Driscoll), Heir Island (artist Tess Leak), and Long Island (artist art manoeuvres). Each art event was connected by a specially commissioned ferry, over an eight-hour journey, which was in turn supported by a unique pedagogical programme, titled the Tidalectic Lecture Series. The Tidalectic Lecture Series brought together artists, islanders, poets, thinkers, educators and government officials to reflect on the changing nature of islands in contemporary society.

The second event was developed over two archipelagos in the global south, the Jambelí Archipelago on the Ecuadorian–Peruvian border and San Cristobál Island in the Galápagos islands in 2019. Supported by Marie Curie 'Excellence' funding, the second iteration focused on the roles of islands in signifying climate change. Central to this iteration was an emphasis on listening and sounding as a relational form which can capture the multi-scalar, geopoetic dynamic of the Anthropocene, which culminated in the Listening School, a conceptual school that was set up in the Galápagos Islands for three days from 24 to 27 July, and linked to West Cork and the Skibbereen Arts Festival in Ireland. Interrogating the Western eco-colonial gaze on islands in the Anthropocene, the Listening School developed a pedagogical programme around listening and sounding the archipelago. Consisting of eight one-tonne sandbags, a series of Listening Pods, each with an audio device buried in the sand, translated the discussions alongside environmental soundings from the different archipelagic regions. Through these geo-locational, aesthetic events, a unique focus on connectivism and digital education began to emerge, influencing the final iteration of the project: The Digital Archipelago (2022).

With global pandemic restrictions at their peak in the winter of 2022, the third and final iteration of the project was realised online in the form of a digital archipelago designed for artistic research and educational exchange between islands, artists and academia. Developed in anticipation of the archipelagic MA in art and environment in the West Cork region, the digital archipelago formed the backdrop to a virtual education programme aimed at supporting environmental art education using Oculus Rift VR headsets. Through this program, students used virtual island spaces to develop contributory research with islanders for artistic proposals on islands after the lockdown. The student's virtual installations captured island sounds and rhythms, wind, birdsong, rustling trees, the sea, and motorboats, and the voices and experiences of islanders and island life during the pandemic. While these experiences provided much-needed novelty and excitement for the students during extreme routinisation and limitation, they also raised important pedagogical questions about embodied knowledge experiences within virtual environments and the need to engage further with the connections between the bio-sphere and the techno-sphere.

Developed over diverse geographic sites from West Cork to the Galápagos Islands, *What Is an Island?* negotiated the shifting relations between locality and world through an educational question, at a time of planetary transition. While it might seem oddly didactic to identify this question as specifically educational, Gert Biesta argues the need to reclaim our understanding of 'educational questions' beyond their normative function in training and instruction. As he writes: 'Educational questions are fundamentally existential questions about how we try to exist as human beings, how we try to live our life well *in* and *with* a world that is not of our making' (Biesta, 2021, p. 9). Importantly for Biesta, the educational work led by such questions is always external to the student, rather than from within the student. This subtle distinction points towards a shift from student-centred learning towards what Biesta calls a 'world-centred education' in his book title which, in the context of the *What Is an Island?* project, can be understood and expanded to infer: Island–Archipelago–World, as an educational event.

Although inspired by an educational question, the *What Is an Island?* project was also motivated by artistic deliberations concerning the form of its inquiry, which was developed through the methodological technique of assemblage, iteration and event. Referencing the work of Jacques Derrida, Barbara Bolt suggests that 'iterability—whether it is in performing language, performing gender or performing art—is the mechanism through which there is movement and transformation' (Bolt, 2016, p. 136). On this basis, she argues that artistic research is a 'performative paradigm', which contributes to the field of research through its capacity to produce 'repetition with *différance*':

> [T]he underlying principle of performativity is iterability, and a priori iterability is subject to the dynamics of *différance.* Thus, good performances, bad performances, playful performances and the excessive performances are all generative of difference. Thought in terms of *différance*, performative research necessarily begins to bud and grow in a disorderly fashion. While operating against the backdrop of convention, re-iteration and citation produce repetition with difference, rather than repetition of the same.
> *(Bolt, 2016, p. 134)*

Similarly, for educator Catherine Ryther, iteration is understood as a productive form in radical pedagogy, supporting an openness to process and spontaneous, situated enquiry, against pedagogical prescription and where 'the unexpectedness of possibility is a crucial feature of iteration. In order to avoid accidentally eliminating the political, a radical pedagogical project of iteration needs to allow for the unexpected and resist adopting end goals for what the imagination of possibility might achieve' (Ryther, 2009, p. 13). While there is a crucial connection between the performativity of iteration and its radical pedagogical form, iteration is also a central geographical and geological dynamic in the formation and reformation of island(s) as archipelagos. Defined by movement, connection and transformation, archipelagoes are geospatial assemblages of becoming that produce and reproduce cultural exchange through creative repetition.

Following this logic, the MA in art and environment was implemented across three islands in Roaringwater Bay in West Cork: Sherkin Island, Cape Clear Island and Whiddy Island. Navigating this chain of Islands, the programme explored three core educational ambitions: (i) to develop a pedagogical ethos that supports research engagements between island communities and academic communities, (ii) to develop an appropriate intellectual framework for navigating environmental change on islands, and (iii) to establish a level 9 educational provision for the study of environmental arts across island communities, locally and internationally. To support these ambitions, a curriculum framework was designed around three dominant modes of governance in the Anthropocene, outlined by David Chandler as Mapping, Sensing, and Hacking (Chandler, 2018).

On the one hand, Chandler identifies each mode as providing a 'distinct conceptualisation of governance in a world framed as complex, entangled and unpredictable' and on the other, each can be mobilised to problematise ontological claims that 'affirm' the Anthropocene (Chandler, 2018, p. 46). Where *mapping* gathers historical and 'empirical knowledge', *sensing* engages with contemporary 'creative assemblages' outside of empirical data 'emphasising an intuitive, experiential response to the complexity of the environment' (Chandler, 2018, p. 48). Finally, Hacking encourages imaginative action and 'new creative ways of engaging on the basis of repurposing, recompositioning and finding the play in already existing arrangements and practices' (Chandler, 2018, p. 49). Similar to William Pinar's disciplinary conception of curriculum, the vertical conception of knowledge concerns the intellectual histories and contexts of a particular enquiry (Mapping), whereas horizontal explorations are informed by circumstantial and experiential learning (Sensing and Hacking), as outlined below.

Engaging with the concept of archipelagraphy across multiple geographical sites, the first module introduces students to the key historical themes, debates and conceptual frameworks that have emerged across island studies and in the environmental arts historically. Developed by island scholar Elizabeth De Loughrey, archipelagraphy is a framework for mapping the historical relation between continents and islands as 'a shifting history between colonial power, economic expansion and environmental change' (De Loughrey, 2001, p. 32). These histories are explored through aesthetic, social and political forms, from the biological and earth sciences, anthropology, visual art, design and political activism. Through these analyses, students are introduced to mapping as a 'diagrammatic methodology' for aiding further fieldwork on human and natural ecologies throughout the West Cork archipelago. These artistic methodologies introduce students to the interdisciplinary and transdisciplinary character of the environmental arts and provide artistic frameworks for the critical mapping of key themes, concepts and questions.

While students are required to 'cognitively map' areas of contemporary environmental art practice and discourse, mapping as an artistic research methodology enables students to build artistic frameworks for participatory fieldwork in the second semester, which shifts its emphasis from mapping to sensing the environment. Sensing fieldwork is essential to an eco-social art practice that engages with human and non-human actors in the environment. Within this context, the programme supports models of contributory research premised on collaboration, creativity and intervention to systematically revalue the acquisition of knowledge through research activities that occur outside of centre/periphery binary definitions of research. In this sense, the incorporation of artistic methods into research is not meant to illustrate island culture merely but participates in the co-production of knowledge with multiple island communities through creative methodologies. Opening up new spaces for dialogical exchange and providing a mechanism for the archipelago to be understood as both a specific and a generalisable geo-historical framework, the sensing module connects the students with local communities and environmental habitats on Islands.

Where students accumulate empirical knowledge in module 1 through cognitive mapping and aesthetic knowledge and in module 2 through experiential, participatory fieldwork; in the third module, they are required to deliver a final project that 'hacks' the local environment. Taking a transdisciplinary approach, the third module encourages novel forms of creative disruption and critical intervention in any system, be it technological, social, economic or environmental. Through this process students are supported to develop large-scale interventions and events in an environmental context. These interventions are further presented in an archipelagic exhibition across multiple island sites with multiple communities and actors.

Conclusion

The *What Is an Island?* project initiated an evental mode of artistic research across multiple island regions to open up new spaces for studying art and environment in situated contexts. Each iteration of the project explored a set of thematics relevant to the study of climate change on islands, beginning with the first project which focused on *mapping* as a pedagogical form that could connect the island communities of West Cork. Following this, the project's second iteration explored islands and archipelagos through the lens of climate change, focussing on *sensing* the islands through mediated listening practices across the global south. The third and final iteration of the project experimented with immersive education for distance learning by *hacking* and repurposing virtual reality platforms for environmental education. While these events were mapped out iteratively across three distinct spatial contexts, their collective impact was to motivate the development of a full-time archipelagic master's programme in art and environment, delivered across the islands of West Cork.

In the first iteration of the MA in art and environment (2022–2023), islands were a central educational form, acting as a concrete reality through which to explore the impact of climate change on island environments, on island communities as models of climate resilience, and on the metaphorical and aesthetic dimension of the archipelago as an organising principle and compositional tool. Whilst significantly interrupted by COVID-19, the students in the course's first iteration developed novel forms of engagement with island communities, both physically and virtually. Moving between these spaces over seventeen months, the students shaped and sculpted their understanding and experience of climate change through performances, eco-acoustic work, virtual reality, and socially engaged art. By grounding their research on the environment, in the local environment and with the most vulnerable communities, their artistic events contribute to our understanding of the event that is climate change. The project's second iteration (2023–2025) began to move away from the landmass cultures of the island and the archipelago towards the fluid depth of the ocean. In this shift, the students turned their attention to wet ontologies and oceanic thinking, developing floating methodologies and aesthetic-scientific experiments with slow violence. Through these experiments, the students began to identify more with Philip Hayward's concept of the aquapelago, defined as 'an assemblage of the marine and land spaces of a group of islands and their adjacent waters' (Hayward, 2012, p. 4).

The MA in art and environment aimed to retain the essential characteristics of archipelagic thinking: openness and connection by transposing the theoretical framework of archipelagic thinking onto the methodological ground of artistic events. In this sense, the course cultivated a response to the event of climate change, providing an assemblage of educational experiences and resources for students to begin to understand the complexity of climate change; its spectacularity and its ambiguity. Constructing an educational form in response to the event of climate change through the archipelago grounds, shapes, and performs the event of climate change for students. These events and performances are iterative and serial and connect the students to alternative forms of representation in the Anthropocene.

Through the tripartite curriculum structure of *mapping, sensing*, and *hacking*, the students are slowly brought through our shifting sensibilities in the post-climate change era. By mapping the environmental events that have historically led to the Anthropocene in Semester 1, they are able to visualise and interpret the geo-political structures that have led us to this point. Moving between islands throughout this study, the student makes connections between the cognitive experience and the local everyday reality of climate change. Working with island communities in semester 2, the students gain a better understanding of the temporalities and scales that inform indigenous understandings of climate change through sensing, aesthetics and ambiguity. At the same time, these structures combine to make climate change 'knowable, visible, audible, recognisable or legible' (Lewis, 2020, p. 18), in the third and final semester the students also create artistic events, which contribute to the passing on or relaying of the event through artistic form, through performances, interventions, sculptures, etc., giving shape again to the 'otherwise unknowable, invisible, silent and so forth' (Lewis, 2020, p. 18).

References

Atkinson, D. 2007. Pedagogy against the State: Some Remarks on Events of Learning. https://pesaagora.com/access-archive-files/ACCESSAV26N2_095.pdf.

Beuys, J. 2006 [1973]. I am Searching for Field Character. In C. Bishop (ed.), *Participation*. Whitechapel & MIT Press.

Biesta, G. 2021. *World Centred Education: A View from the Present*. Routledge.

Bolt, B. 2016. Artistic Research: A Performative Paradigm. *Parse Journal*, vol. 3, pp. 129–142.

Britzman, D. 2000. If the Story Cannot End: Deferred Action, Ambivalence, and Difficult Knowledge. In R. I. Simon, S. Rosenberg, and C. Eppert (eds), *Between Hope and Despair: Pedagogy and the Remembrance of Historical Trauma*, pp. 27–57. Rowman & Littlefield Publishers.

Chandler, D. 2018. *Ontopolitics in the Anthropocene. An Introduction to Mapping, Sensing and Hacking*. Routledge.

De Loughrey, E. 2001. 'The Litany of Islands, the Rosary of the Archipelagoes': Caribbean and Pacific Archipelagraphy. *ARIEL*, vol. 32, pp. 21–51.

Derrida, J. 2011. *The Beast and the Sovereign, Vol. 2*. University of Chicago Press.

Dimock, W. C. 2006. *Through Other Continents: American Literature Across Deep Time*. Princeton University Press.

Friedman, S. S. 2015. *Planetary Modernisms*. Columbia University Press.

Glissant, É. 1997. *Poetics of Relation*. Translated by B. Wing. University of Michigan Press.

Glissant, É. 2009. *Philosophy of the Relation: Poetry in Extension*. Gallimard.

Hayward, P. 2012. Aquapelagos and Aquapelagic Assemblages. *Shima: The International Journal of Research into Island Cultures*, vol. 6, no. 1, pp. 1–10.

Heise, U. 2008. *Sense of Place and Sense of Planet: The Environmental Imagination of the Global*. Oxford University Press.

Lee Podevsa, K. 2007. Notes on the Pedagogical Turn. https://designopendata.wordpress.com/portfolio/a-pedagogical-turn-brief-notes-on-education-as-art-2007-kristina-lee-podesva/.

Lewis, T. 2020. *Walter Benjamin's Antifascist Education: From Riddles to Radio*. SUNY Press.

Loughran, G. 2020. Archipelagic Imaginaries: A World-Centred Art Education at the End of the World. https://doi.org/10.13137/1825-5167/30983.

McKenzie, I. and Porter, R. 2016. Evental Approaches to the Study of Events. In T. Pernecky (ed.), *Approaches and Methods in Event Studies*. Routledge.

McCall, G. 1994. Nissology: The Study of Islands. *Journal of the Pacific Society*, vol. 17, pp. 1–14.

Nixon, R. 2011. *Slow Violence and the Environmentalism of the Poor*. Harvard University Press.

Rancière, J. 2013. *The Politics of Aesthetics*. Bloomsbury Academic.

Ryther, C. 2009. The Possibility of Living Another Life: Reimagining Possibility in Radical Pedagogy through Rancière. www.academia.edu/1291209/_The_Possibility_of_Living_Another_Life_Reimagining_Possibility_in_Radical_Pedagogy_Through_Ranci%C3%A8re (accessed 27 August 2024).

Sewell, W. H. 2005. *Logics of History. Social Theory and Social Transformation*. Chicago University Press.

Shlensky, L. Z. 2013. Édouard Glissant: Creolization and the Event. *Callaloo*, vol. 36, no. 2, p. 356.

Strayer, E. 2019. Recentering the Peripheral: An Event-Based Ecocritical Methodology for World Literature. *New Global Studies*, vol. 13.
Tavory, I. and Wagner-Pacifici, R. 2022. Climate Change as an Event. *Poetics*, vol. 93, part A, August, 101600.

6

TEACHING ECOCRITICAL ART HISTORY

Timothy Stott

Introduction

How might we teach ecocritical art history? Art historians feel the need, like other humanists, to recast the natural world as a matter of concern, but the question of how this might change the teaching of art history receives less attention. Whatever innovations art history might now make in content and methodology—and there are several and they are vital—as the discipline rather belatedly joins the environmental humanities, there remain more prosaic, but no less urgent, challenges concerning curriculum, assessment, and pedagogy, so that those innovations translate into education. Our climate futures loom large over a discipline that is often quite slow and conservative, often gratefully at some remove from the knowledge economy and the news cycle. The historical strengths of art history can jar with students' sense of crisis or fade under the clamour of solutionism, where other disciplines and other faculties more readily provide answers. Alan Braddock and Christoph Irmscher have rightly noted that the challenge of teaching ecocritical art history exceeds the matter of how to integrate a history of 'overtly engaged or sympathetic forms of environmentalist expression' into existing curricula, as an additional field, say. 'Every work of art has environmental significance and is therefore open to ecocritical inquiry', they assert (Braddock and Irmscher, 2009, p. 9). I would argue that an ecocritical art history also requires a more fundamental rethink of how the discipline is taught, how it brings the world into the classroom and how it meets the world outside. An ecocritical art history has emerged rather piecemeal over the past couple of decades or so. For Braddock, writing fifteen years ago, a historically oriented ecocriticism 'may bring attention to neglected evidence of past ecological and proto-ecological sensibility or it may cast canonical works and figures in a new light by

DOI: 10.4324/9781003456636-7

revealing previously unnoticed complexity regarding environmental concerns' (Braddock, 2009, p. 26). More recently, Andrew Patrizio proposes an ecocritical art history assembled around political commitments to non-hierarchy and philosophical commitments to so-called 'flat ontologies', which would decentre humans, study the roles assigned to non-humans in service to the artistic enterprise, and connect with the pursuit of climate justice outwith the university. For Patrizio, ecocritical art history would 'explore a sensory attentiveness to the nonhuman' (Patrizio, 2019, p. 20). This would require a close engagement with the natural sciences, of course, but also with the institutional consequences of any commitment to nonhierarchy. Therefore, as the field debates its commitments and orientation, we should remember that, as Terry Eagleton once remarked, in the university (and especially in the humanities), the most radical ideas tend to be taught in the most conservative of ways. We also tend to promote our discipline's environmental awareness and responsibility without discussing the hidden curriculum of the campuses and other institutional spaces in which we teach.

This is why we also need to discuss teaching. Colleagues in the humanities find as they attempt to turn their teaching toward the climate emergency that the best intentions and the most sophisticated proposals still run aground against the long arc of institutional change, disciplinary bias, teaching habits, and a hierarchy of urgency over importance and of applied over basic research. The latter is especially difficult to circumvent. Solutionism, technoscience, and markets dominate responses to the climate emergency. Added to this, whatever pedagogical reforms ecocritical art history requires must, of course, negotiate institutional differences in curriculum design, in university strategies and priorities, in disciplinary histories, in local and national funding structures, and in the legacies of previous conflicts of the faculties that still divide the university. In Ireland, a country that has been lethargic in facing its climate challenges, the evolution of ecocritical art history has been stunted by a funding and institutional landscape that greatly overvalues applied research at the expense of basic research on environmental topics and an art history that has tended toward the connoisseurial and biographical and that has been slow to rethink materials, agents, artefacts, and ecologies in light of broader posthumanist debates in the humanities. As a result, Irish art history remains one of the most humanist of humanities disciplines. A lingering distrust of 'theory' follows from the misconception that it is at odds with rigour, evidence, and judgement. However, to borrow from Stefan Collini, theory 'is what happens when common starting points can no longer be taken for granted' (Collini, 2012, p. 69). This latter is our situation as scholars and teachers in universities.

We need to theorise our teaching of ecocritical art history based upon the demands of the classroom and the strengths and blind spots particular to the subject. To this end, I offer some reflections on my experience of trying to teach ecocritical art history over the past few years at Trinity College Dublin. I can boast neither strategy nor experiment. The lectures and modules that I have taught worked within the possibilities of the current curriculum. Nor do I pretend to offer any unique pedagogy or model of best practice. Nevertheless, ecocritical art history remains a field very much in formation *in the classroom*, which means that sharing experience of teaching, its challenges and possibilities, is vital. This is the spirit in which the following should be read.

Possibilities and Challenges

First, let me reassert the necessity and challenge of an ecocritical art history. Amitav Ghosh (2016) argues that we still struggle to imagine climate change, its causes and its consequences, not because we lack the science, of course, but because such change remains stubbornly improbable, even unimaginable, in the stories we tell ourselves about the natural world. The humanities have some responsibility here, conceiving of the natural world as the background to the great human dramas they study. Ghosh, a novelist and essayist, writes of literature, but his critique applies equally to art history. Nature appears in the artefacts art historians study mostly as source, inspiration, model, setting, and so on, but rarely if ever prone to systemic or catastrophic change and rarely if ever outside the domain of human interest. We now know that human drama and planetary history cannot separate so easily into background and foreground. As Dipesh Chakrabarty has famously argued, 'anthropogenic explanations of climate change spell the collapse of the age-old humanist distinction between natural history and human history', which brings the planet and its actors into the foreground of human drama and of the humanities (Chakrabarty, 2009, p. 201).

What this means for the history of art is still a matter of debate. How can art historians think across spatial and temporal scales, from a planetary 'shared sense of catastrophe' to more local historical and cultural contingencies, as Chakrabarty (2009, p. 222) urges? How might art history develop an ecological perspective on its artefacts? This would be a history of artworks that took the natural world as their theme, such as an ecological history of landscape art that studies how aesthetic sensibility and ecology may have converged, or a history of painting's role in framing nature for purposes of settlement, reclamation, extraction, colonisation, and so on, however important these studies undoubtedly are. It would also be an art history where the natural world is brought to the foreground as condition of the wealth of patrons, collectors, and artists; as source of materials and energy as much as motifs and scenes; as historical actor and as measure of what is meaningful and what is valued. Art historians of earlier periods might learn here from those developments in the history of modern and contemporary art that have corresponded to its expanded field of practice, its proliferation of media, its responses to environmental crisis, and its attempts to conceptualise the environments of art (De Bruyn, 2019; Nisbet, 2014). Exemplary in this regard would be Sugata Ray's *Climate Change and The Art of Devotion: Geoaesthetics in the Land of Krishna, 1550–1850*, published in 2019. Ray applies geoaesthetics, a term coined to describe land art and environmental art of the 1970s and after, to the correspondence of environment and devotional practices that emerged in the Braj region in the early modern period. This correspondence of ecology and aesthetics is disclosed, Ray demonstrates, through that region's art and architecture. Geoaesthetics, Ray claims, 'unmoors art history's epistemological scope' so that the human is no longer its primary conceptual category (Ray, 2019, p. 23).

Michael Baxandall argued over forty years ago that one of the main pleasures of art history was its 'excursions into alien sensibilities', to gain 'at least some intimation of another culture's visual experience and disposition' (Baxandall, 1982, p. 143). An ecocritical art history would extend these excursions, analysing historical artefacts as documents of how other cultures experienced, imagined, valued, and behaved toward natural environments and the organisms that inhabited them, but would also need to investigate non-human sensibilities and their ecologies if it is to challenge what Braddock and Irmscher (2009, p. 5) call 'the entrenched speciesism, or fundamental anthropocentrism, that has long governed this and other humanistic disciplines'. Such are some of the possibilities and challenges of an ecocritical art history. Guidance in meeting these challenges can be found in recommendations for teaching across the environmental humanities. As one would expect from an interdisciplinary field in formation, there is little consensus about what the environmental humanities mean, what they include, or how they should be taught (see Heise et al., 2017; Emmett and Nye, 2017). In addition, broad recognition of their value across the university or within the political and professional institutions tasked with tackling the climate emergency remains quite remote. Nevertheless, this is a growing and vital field of humanities scholarship, which pushes up against the constraints of the current curriculum. The need for pedagogical experiment is clear. An international survey of environmental humanities teaching, published in 2019, provides several examples of such experiments. Many promote 'grounded thinking' through case studies and through insisting that students situate their knowledge in relation to other disciplines and to the world (O'Gorman et al., 2019). Because of the latter, many also understand that environmental humanities teaching must take the world as its classroom. This is where I wish to return to my own teaching.

Towards the World as Classroom

To repeat, at its most basic, teaching history of art as one of the environmental humanities means to show students that cultural artefacts can be studied 'as if nature mattered' and as if their concerns about the planet mattered. It is vital that students learn to match their programme of study to their concerns about the climate emergency, to learn how they might live in and with a world in crisis. Most tend to carry into their third-level education an assumption that the climate emergency is not the concern of the humanities. If we teachers do not correct this assumption, students can experience a lack of agency or consequence and fall into cynical reason. Informal conversations with my undergraduates, many of whom would have joined the Fridays for Future demonstrations beginning in 2018, have revealed their frustration with political inaction and half-measures and a deep distrust of market-driven sustainability, coupled with, at times, embarrassed reflections on the political demands and commitments of their adolescence. There is often a pathos and a political charge to these classes and seminars, which places a responsibility upon the teacher to add the need to act to the familiar and necessary effort to understand and to explain. The teacher becomes one who *should* know *how to act*. In my experience, this is a rare and humbling responsibility for a professor of art history to have. The urgency of this demand also makes it difficult to stay in and with the complexities of the world.

In many respects, art history already takes the world as a classroom when it teaches from collections, archives, buildings and spaces, visual and material culture, and so on. Clearly, these artefacts are very much of the world. In other words, teaching art history is already committed, in part, to what Gert Biesta calls a 'world-centred education'. But this latter requires more than exiting the classroom. If we follow Biesta, teaching orients students toward 'being at home in the world'—a phrase Biesta borrows from Hannah Arendt's 'Understanding and Politics (The Difficulties of Understanding)' from 1953—and encourages them to strive toward 'grown-up-ness' in a world that is, we should add, acutely uncertain and threatened. Biesta notes that this requires a teacher to interrupt students' desires, bringing the question of what is being expressed [by a student], or, in other words, the desirability of what is being desired, into the lives of students, *not* as a question that can be answered once and for all, but as a question that is worth asking in each new situation, as a question that is worth carrying with oneself throughout one's life (Biesta, 2018, p. 149).

Such interruption allows teachers and students to identify 'which desires are going to help in our attempts at existing in and with the world in a grown-up way, and which desires are going to hinder us in that task' (Biesta, 2018, pp. 149–150). At the core of Biesta's views on education is the claim that the world, rather than simply ourselves, provides the criterion for our existence. The point of education is to find the world's 'point of integrity', a difficult balance between our desires that end with 'world-destruction' or with 'self-destruction' (Biesta, 2023, p. 16). Rarely has finding this integrity been more important to teachers and students. To centre education on either the learner or the curriculum is to divest from those ways in which education helps us to meet the natural, social, and physical world. 'This is perhaps the first question of the curriculum—not in terms of what children and young people should 'learn' or 'get' from the curriculum, but what they can meet and encounter there' (Biesta, 2023, p. 16).

Biesta's promotion of world-centred education finds an ally in the teaching of art history. The latter is interruptive because a student's intentions and interpretations come up against the materials and complex lives of artefacts from historic worlds in a situation where artefacts answer back to (and sometimes against) the desire to classify, understand, and interpret. The artifact can tell us, of course, about how others' desires, thoughts, experiences, values, and disposition have met worlds that our not our own but whose legacy we experience. There is an intelligence and sensibility to these artefacts, the discovery of which requires students to check their rush to interpret according to their own assumptions and desire for meaning (see Grootenboer, 2020; Crow, 1999). In our design of the curriculum, our primary expectation for learning is that students encounter past worlds through artefacts and that they consider the legacy of those worlds in the present. Any concerns with learner attributes are subordinate to these encounters. In fact, a principal feature of our first- and second-year undergraduate teaching is to reorient students from the learner-centred approach used during their secondary education toward a world-centred education at third level. The historic worlds that students discover through an ecocritical art history would include more than human actors in the production of artefacts and would assess past human actions and their own capacity for action in light of these other actors' histories. Students might well discover that world-destruction and artefact production often went hand in hand. Understanding this interrupts students' assumptions about the value of art and would be central to a grown-up and grounded study of art history.

The Future Grows Big

Although a historical discipline, the challenges of teaching an ecocritical art history are not wholly met through encounters with historic artefacts. A further challenge, felt keenly by those trying to connect the study of historical subjects to students' concerns about the future, is that, to paraphrase Chakrabarty, contrary to the assumptions of most historical scholarship, we cannot envisage the future in the same way we imagine and seek to reconstruct the past. The world in and with which we might teach students to be at home is divided. In the undergraduate open module HHU22003 Histories and Futures of Climate Change, co-taught with my colleague, the environmental historian Dr Katja Bruisch, we present this problem to students and try to guide them through it. Support from a Mobility Grant from the Higher Education Authority has enabled us to integrate study trips, such as to the Centre for Sustainability, Equality, and Climate Action (SECA) at Queens University or to the National Botanic Gardens in Dublin, both of which demonstrated to students the possibilities of interdisciplinary research on climate in situ. After an afternoon of workshops with PhD researchers at SECA last year, many students took a much more political approach to their own subjects. 'We are all Marxists now', one quipped afterwards. In their study groups, some then focussed on flood risk in Belfast as a consequence of architecture, planning, agriculture, conflict, and the complex psychology of land. Others compared Belfast to flood risk Dublin suburbs such as Clontarf and Sandymount. One group completed an impromptu survey of Belfast shoppers, asking them how they imagined climate change.

This year, study groups took Dublin city centre or the Trinity campus as case studies, assessing their mitigation and adaptation strategies to climate change. The module is only a couple of years old. It is exciting to teach and seems to be rewarding for students, but it is still very much a work in progress, both in its syllabus and its pedagogy. Students' most encouraging feedback has been that they can see how to match their subject with their concerns about the climate emergency and with the places they inhabit as students and citizens. As David Orr (1992) noted over thirty years ago, learning ecological literacy must be grounded in a subject and a place, but the complexity of the latter shows students both the necessity and limitations of their respective disciplines. This has become especially clear to us over the past two years teaching HHU22003. This is only a first step, of course. The case studies, for example, are vital for these student groups to make the world their classroom. Katja and I need to experiment further to transform them from multidisciplinary to interdisciplinary. It is a tall ask of Senior Fresh students to succeed where many scholars fail, but at best they approach their disciplines with an informality and generosity that we hope to foster. They also become aware at quite an early stage in their studies of the knowledge cultures of their and their peers' subjects. The art history components of this module include a lecture on visualising climate, which discusses artworks but also climate models, information design, remote sensing infrastructures, and disaster films. Another lecture is Designing Climates, which surveys twentieth-century architecture and design for attempts to control and, more recently, reconstruct environments and habitats. A third lecture looks to future studies and compares the ways in which the future has been modelled, predicted, and visualised through climate science with the same in experimental art and design practice. These provide students with perspectives on art and design that they have not yet encountered, even those studying History of Art and Architecture, but they also return the whole class to that challenge of how to connect histories to futures, how to orient historical disciplines and the histories of disciplines toward a future that looms large and threatening and for which there seem to be few historical analogues. My aim is that students from several disciplines become aware of how art and design has imagined alternative futures.

The above clearly indicates the difficulty and necessity of integrating what Zoltán Boldizsár Simon and Marek Tamm call 'historical futures' into the teaching of ecocritical art history. As they note, today's challenges increasingly lie with the future. The shadow cast over our present by the times to come has grown so immense in its scale and scope that we struggle to comprehend the predicament in which we find ourselves (Simon and Tamm, 2021, p. 3). Unique, perhaps, for the discipline of history of art, an ecocritical art history feels the pressure of the future upon it and must reassess what value accrues to learning about the past of art without falling into the fallacy of presentism. More recently, Simon writes, As the future grows big and unfathomable, there is an existential question that the historically oriented human and social sciences typically avoid asking … *does the past still matter*? (Simon, 2023, p. 45). I sympathise with Simon's concern but would prefer to ask which and whose pasts still matter, and how, as we seek to anticipate the future.

In such a tumultuous and divided world, the history of art cannot be simply developmental, linking the present to the past through a causal chain. Since the discipline abandoned its Hegelian legacy, the history of art has challenged this developmental approach, not least in the proposal that historical artefacts are anachronic, or belong to multiple times. As Alexander Nagel and Christopher Wood write, 'No device more effectively generates the effect of a doubling or bending of time than the work of art, a strange kind of event whose relation to time is plural' (Nagel and Wood, 2010, p. 9). This applies to 'buildings, paintings, prints, drawings, sculptures, medals, pavements, mosaics' (Nagel and Wood, 2010, p. 13). This plural relation to time means that anachronic artefacts might disclose past futures and might be haunted by all manner of survivals and returns. Hence the renewed interest in the cultural history proposed by Aby Warburg. As Georges Didi-Huberman asserted, 'It is for art history to establish "its own theory of evolution", its own theory of time … other than habitual chronologies, eternal "influences", [or] old Vasarian or neo-Vasarian family myths' (Didi-Huberman, 2002, p. 61). Art history's study of anachrony and survival (or *Nachleben*) connects the past and future in ways that do not simply reduce the latter to what is known of the former. If, as Simon (2023, p. 46) contends, to see the future as either a recurrence or an unfolding of the past is to deny the novelty of the way in which the future now bears upon our historical disciplines, teaching students the historiography of anachronic artefacts might better prepare them to anticipate the future, to understand those artefacts as still potentially oriented to the future rather than relegated to a remote past.

One way in which an ecocritical art history can bring this unique historiography into the places of teaching is through discussion of ancestral futures (Krenak, 2024). This is a vital area of current art-historical research that is ecocritical and anachronic to its core. Lakota activist and scholar Nick Estes (2019) writes that 'our history is the future', in other words, the history of Indigenous peoples subjected to colonisation, enclosure of common lands, extraction of resources, terraforming, and racialised politics, will be the future for many other groups as the environmental crisis unfolds. The future of art history could be ancestral if it understands how the past inhabits our places and our artefacts and folds the study of living histories into the study of alternative futures. As historians, we might learn to ask not when but *where* did something happen, where was something made, and which agents, human and more than human, made it and might be remembered through it. This would align an ecocritical art history with Amitav Ghosh's call for a vitalist politics that recognises the role of the Earth in making a people and their complex, often hybrid and cosmopolitan culture (Ghosh, 2021). Such vitalism is hardly unfamiliar to historians of art, given the peculiar temporality of the artefacts we study and the strangeness of the meanings and agency attributed to them. In my experience of teaching, our students, too, readily accept the proposal of ancestral futures.

Slowing Down

In conclusion, I wish to make a case for slowness as a central value of teaching ecocritical art history. Again, we can turn to Biesta's world-centred education here. In addition to interruption (discussed above), he promotes 'suspension' and 'sustenance', giving time to students to work through their encounters with the world and giving them the support to 'stay in the often difficult "middle ground"' (Biesta, 2023, p. 17). Art history, of course, has a long tradition of slow or close looking. The challenge is to orient such looking toward ecological literacy as a product of visual and spatial literacy. The above is tested, to some extent, in my Senior Sophister seminar HAU44017/8 Art, Design, and Nature since 1930. An impossibly grand title but broad enough to allow for some flexibility, both in the classroom and in the syllabus year on year. The seminar format, two hours per week over two Terms, allows students time to slow down and look closely at a range of primary materials, sites, and places. The aim is to study how nature has appeared in art and design of the twentieth and early twenty-first centuries, and how natural processes have been mimicked, represented, tested, maintained, misinterpreted, and restored.

The combination of art and design allows our discussions to range from abstract sculpture to wildlife habitat design to technical images to architectural exhibitions, which flattens and expands the objects of our attention. '[One] kind of corrective to dogma is looking itself, pursued long enough', Richard Wollheim argued in *Painting as an Art* (Wollheim, 1987, p. 6). Certainly, Wollheim's claim still holds true, but the dogma in question is not simply ready-made interpretations of iconography or style. It is the assumption that paintings, along with all the other artefacts and environments of art, concern only human experience and intention, an assumption bolstered by the social constructivism that has come to dominate the humanities' understanding of the world outside. No longer is it enough simply to slow down to 'coax a picture into life', as Wollheim has it. The question becomes how life was coaxed into the service of the picture and what the picture might still tell us of that life. Close looking 'pursued long enough' here departs from its formalist origins to become 'meticulous observation', which, to borrow from literary scholar Peter Barry, abandons social constructivism and linguistic determinism, fosters a sense of collective responsibility and agency, and attends to 'the claims of the world beyond ourselves' (Barry, 2009, p. 264). An ecocritical art history would commit, necessarily, to world-centred, 'slow' scholarship (see Berg and Seeber, 2016). Its educational goal would be not to prepare a student to act on a world that should meet that student's desires and ambitions but to learn how to be in and with the world, to meet its complexities slowly and up close. (One wonders how an ecocritical art history committed to world-centred education will negotiate the pedagogical assumptions of the UNESCO Education for Sustainable Development that is now been tested in many universities. ESD promotes constructivism and PBL, places the learner, rather than the world, at its centre, and recasts scholars and teacher as facilitators.)

This is not to withdraw art history from the pressing demands of the present. It is, rather, to assert its value. In my view, an ecocritical art history should be taught across humanist and scientific disciplines through a curriculum flexible, imaginative, and bold enough to allow some interdisciplinary undergraduate teaching. Its ambition would be, for example, the study of those images and designs that have shaped our understanding of nature past and present and whose legacy, for better or worse, prepares our future. It would be a *Bildwissenschaft* or even a *Gestaltungwissenschaft* across the university. Ecocritical art history can show that images and designs have been and remain technical, epistemic, and normative. They have combined what we know with what we value. In doing so, ecocritical art history shows the importance of art-historical methods to analysing a much broader array of materials than just art. Consider, for example, Sheila Jasanoff on the role of images in scaling our imagination to a global environment. Whether in the history of science or in the history of art, it seems that images become persuasive only when ways of looking at them have been carefully prepared for in advance, through the creation of a stylised visual idiom or an interpretative tradition that knows how to respond to particular types of images. The meaning of pictures is inseparable from the context that supplies the idioms of interpretation (Jasanoff, 2001, p. 318). The study of 'stylised visual idioms' and 'interpretative traditions' is long familiar to art historians, and they have been adept at expanding the variety of artefacts for study and showing the importance of their discipline to understanding the range of human experience and imagination. My hope is that an ecocritical art history can push the discipline further into alien domains. The messy, difficult, sometimes frustrating, sometimes astonishing work of teaching is central to this test of art history's worldliness.

References

Arendt, H. (1953). Understanding and Politics (The Difficulties of Understanding). *Partisan Review* XX (4): 377–392.

Barry, P. (2009). *Beginning Theory: An Introduction to Literary and Cultural Theory. Third edition*. Manchester: Manchester University Press.
Baxandall, M. (1982). *The Limewood Sculptors of Renaissance Germany*. New Haven, CT: Yale University Press.
Berg, M. and Seeber, B. K. (2016). *The Slow Professor: Challenging the Culture of Speed in the Academy*. Toronto: University of Toronto Press.
Biesta, G. (2018). Walking the Museum. In C. Naughton, G. Biesta, and D. R. Cole (eds), *Art, Artists, and Pedagogy: Philosophy and the Arts in Education*. London: Routledge.
Biesta, G. (2023). Putting the world in the centre: A different future for Scotland's education. *Scottish Educational Review* 54 (2): 149–169.
Braddock, A. C. (2009). Ecocritical Art History. *American Art* 23 (2): 24–28.
Braddock, A. C. and Irmscher, C. (2009). Introduction. In A. C. Braddock and C. Irmscher (eds), *A Keener Perception: Ecocritical Studies in American Art History*. Tuscaloosa, AL: University of Alabama Press.
Chakrabarty, D. (2009). The Climate of History: Four Theses. *Critical Inquiry* 35 (2): 197–222.
Collini, S. (2012). *What Are Universities For?* London: Penguin.
Crow, T. (1999). *The Intelligence of Art*. Chapel Hill, NC: University of North Carolina Press.
De Bruyn, E. C. H. (2019). A Proposal: Must We Ecologise? *Grey Room* 77: 58–65.
Didi-Huberman, G. (2002). The Surviving Image: Aby Warburg and Tylorian Anthropology. *Oxford Art Journal* 25 (1): 59–69.
Emmett R. S. and Nye, D. E. (eds). (2017). *The Environmental Humanities: A Critical Introduction*, Cambridge, MA: The MIT Press.
Estes, N. (2019). *Our History Is the Future: Standing Rock versus the Dakota Access Pipeline, and the Long Tradition of Indigenous Resistance*. London: Verso.
Ghosh, A. (2016). *The Great Derangement: Climate Change and the Unthinkable*. Chicago, IL: University of Chicago Press.
Ghosh, A. (2021). *The Nutmeg's Curse: Parables for a Planet in Crisis*. Chicago, IL: University of Chicago Press.
Grootenboer, H. (2020). *The Pensive Image: Art as a Form of Thinking*. Chicago, IL: University of Chicago Press.
Heise, U. K., Christensen, J. and Niemann, M. (eds). (2017). *The Routledge Companion to the Environmental Humanities*, London: Routledge.
Jasanoff, S. (2001). Image and Imagination: The Formation of Global Environmental Consciousness. In C. A. Miller and P. N. Edwards (eds), *Changing the Atmosphere: Expert Knowledge and Environmental Governance*. Cambridge, MA.: The MIT Press.
Krenak, A. (2024). *Ancestral Future*. London: Polity Press.
Nagel, A. and Wood, C. (2010). *Anachronic Renaissance*. New York: Zone Books.
Nisbet, J. (2014). *Ecologies, Environments and Energy Systems in Art of the 1960 and 1970s*. Cambridge, MA: The MIT Press.
O'Gorman, E.*et al.* (2019). Teaching the Environmental Humanities: International Perspectives and Practices. *Environmental Humanities* 11 (2): 427–459.
Orr, D. W. (1992). *Ecological Literacy: Education and the Transition to a Postmodern World*. Albany, NY: State University of New York Press.
Patrizio, A. (2019). *The Ecological Eye: Assembling an Ecocritical Art History*. Manchester: Manchester University Press.
Ray, S. (2019). *Climate Change and the Art of Devotion: Geoaesthetics in the Land of Krishna, 1550–1850*. Seattle, WA: University of Washington Press.

Simon, Z. B. (2023). Does the Past Still Matter? In P. Dover (ed.), *Engaging with the Past and Present: The Relationship between Past and Present across the Disciplines.* London: Routledge.
Simon, Z. B. and Tamm, M. (2021). Historical Futures. *History and Theory* 60 (1): 3–22.
Wollheim, R. (1987). *Painting as an Art.* London: Thames & Hudson.

7

LETTING THE DEAD TEACH

The Pedagogy of Arkadi Zaides's *Necropolis*

Juliette Clara Bertoldo

Introduction

What can the dead teach the living in and for a post-human-capital world, and how? This is the key question addressed in this chapter. To formulate an answer, I transpose the model of an ethico-political interaction with the dead that choreographer Arkadi Zaides proposes in his research-based performance *Necropolis* (2017) in order to draw out its pedagogical dimension. Charting a path from the movement to Jacques Derrida's (1994) hauntology to Vinciane Despret's (2023) methodology for 'making memory *with* the dead' (p. 16), Zaides's work provides pedagogical avenues for ethical engagement with the unjustifiably dead, aimed at drawing the living toward what Delphi Carstens (2021) calls 'a justice-to-come'—that is, 'a more inclusive sense of justice that resists the gendered, racist and anthropocentric onto-theological legacies of capitalism' (p. 124). In seeking to move beyond problematic modes of representation of forced illegal migration (Banerjee et al., 2023; Gallien, 2018), *Necropolis* convincingly suggests that such an approach to a posthuman justice requires practical and conceptual tools to enable the living to encounter the 'other' through their death, as a premise for an ethical response that would allow us (the living / the public) to learn 'to live differently', i.e. 'more justly' and 'with them', in the words of Derrida (1994, p. xviii).

In this, the staging of an aesthetic encounter (Todd, 2023) between worlds strictly separated by Western onto-epistemic normative orders presupposes recognition of the fundamental principle that death does not dissolve the relational obligation of the living to these other beings (see, for example, Despret, 2021; Lykke, 2022; Shildrick, 2020). Here, these others are not just any dead, but those who have found themselves outside the protection of the nation due to the political crises associated with migration (and exacerbated

DOI: 10.4324/9781003456636-8

by current anthropocenic conditions);[1] they are the 'excluded, expelled and invisible others of social production' who have become, as Carstens (2021) puts it, 'the ghosts of those bodies not included in the category of "human"' (p. 123). Zaides's art turns to them, inspiring new imaginings of death for a post-capital, justice-oriented pedagogy that stirs strange hopes for the deceased victims of displacement and exclusion. While examples of artistic projects on migration injustice abound, with work such as *The Jungle* by Robertson and Murphy (2017), *Law of the Journey* by Weiwei (2017), and *Lampedusa* by Lustgarten (2015), to name but a few, Zaides's powerful and unique proposal reveals that migrant subjects denied rights in life are denied rights after life, in their death. Rejected by hegemonic powers and jurisdictions, the process of othering to which migrants are subjected during their exile is prolonged in death, excluded from 'grievability', in Judith Butler's sense (Butler, 2004, pp. xiv–xv), an exclusion that maintains and produces a hierarchy between the dead, i.e. those who deserve public recognition and those ignored, often anonymous and conveniently forgotten. This denial of death and justice echoes Despret's (2021) provocative assertion that abandoning the dead is tantamount to killing them a second time. In refusing this redoubling of death, *Necropolis* is an important response to racist violence, which the artist expresses through the creation of an ethico-political and artistic project of posthuman sensibility in its attempt to develop new relationships with the nonhuman, 'with those who have no voice or cannot be seen', and to imagine other potentialities of existence (jagodzinski, 2015, p. 126), through its ability to combine forensic science, art practice, and activism. Pedagogically, as we shall see, the aesthetic encounter forged by Zaides offers an occasion to connect subjectivities beyond the spatio-temporality of the performance itself, where the audience is not simply an observer, but brought into relation of response.

This essay begins by defining Despret's parameters and Derrida's conceptual ideas, which are then mobilised for my reading of *Necropolis*. I then provide a description of the work integrated within a discussion of the tools in use that include counter-mapping and the reenactment of a forensic ritual. Through this pedagogical analysis, divided into two parts, I seek to highlight the way in which art lets the dead speak and thereby teaches us something otherwise silenced and invisibilised. This educative force is facilitated by the modes of resistance and gestures of care staged by the artist, which consist not only in making us see what is invisible, but also in seeing what is not there, as a deeply felt form of haunting confronting us with what is in the present as ongoing violence. I will conclude with a brief summary on the pedagogical possibilities offered by this artwork for thinking differently about spatio-temporality, aesthetics, and political action in the service of a 'liveable and more inclusive future' amidst the ruins of capitalism and environmental racism (Carstens, 2021, p. 121).

The Haunting Ghost and the Dead as Subject

Both Derrida and Despret are theorists who have productively placed the dead within our thinking. The difference between them is an important and productive one. In *Specters of Marx* (1994), Derrida develops the concept of hauntology, a near-homonym to ontology, i.e. the 'state of being', to give attention to the state of 'non-being' (Derrida, 1994, p. 12)—namely, that which is absent or made ostensibly liminal and yet lingers in the present as a 'haunting' (Derrida, 1994, p. 10). Haunting is a mode of critical, embodied, and dynamic relating to ghosts of past injustices that return unceasingly to haunt the present, in the shadow of forgotten pasts and tragic futures of repetition. If communist spectres are Derrida's central subject, haunting is not limited to Marx's socio-political project, but structures all hegemonic power that have set in motion an unprecedented process of planetary instability, thus manufacturing new spectres (Carstens, 2021). Derrida thus opens up fruitful conceptual avenues for demonstrating that injustices haunt the living, inscribing an unbearable past at the heart of the present, and enables the unveiling of omissions and deliberate erasures so as to account for the ongoing violence of the crosspollinating 'colonial/modern' project (Mignolo, 2011). In contrast, refusing the spectral grandeur of the Derridean approach, Despret (2021) considers the dead as real, social beings, not aspects of the collective unconscious or one's psyche resulting from a repressed injustice or tragedy. She points to the (pedagogical) risk of the spectral as being a 'mortifying trope that comes along to efface the bodies and voices of those who continue to live and pay the consequences of past injustices' (Despret, 2021, p. 54). If she rejects the register of spectrality, it is partly to avoid falling into allegorical modes of representation that only seem to 'express the anxieties and taboos of the epoch' (Despret, 2021, p. 53) with the added harm of trapping singular beings exclusively in relation to retribution, revenge, and victimisation. Instead, Despret focuses empirically on the active, ongoing relations unfolding between the living and the dead, and, through their dynamic exchange, how the latter change the lives of those responding.

To this first distinction, another comes to view. Whereas Derrida in a way states what is the case (i.e. the resonance of those no longer with us that we must learn to live with), Despret articulates a methodology for doing so. Taking seriously the assertion that those who are no longer with us *are* with us and have an impact on us—as Derrida does—she takes this idea a step further: It is the dead who makes the living act. At the heart of her methodology lies the key question: What can the dead make the living do? Despret's (2021) method aims to retrace the stories of 'those who insist' (Despret, 2021, p. 13) on making the living continue what has been left unfinished, thus enabling the dead to accomplish something on their behalf. This prolonged existence can be materialised through the vector of an artwork that consists in 'making memory-*with*' the dead (Despret, 2023; emphasis in original). She writes: 'The dead are endowed with the power to continue to act in this world, not only by helping the living to "do with this world" but also by transforming it through the vector of a work of art' (Despret, 2021, p. 13; my translation). Put simply, the dead are both summoned into the artwork; at the same time, it is they who produce the very fact that there is an artwork.

Zaides plays with both registers. By proposing a pedagogical and performative model oriented toward aesthetic, ethico-political modalities whose aim is to account for and listen to the victims of racial, murderous injustice, he stages a haunting to explore the unspoken that surfaces as unresolved and elaborates new, exciting, dynamic ways of '*re*-presenting' the dead (Despret, 2021, p. 18; my emphasis)—namely, to make the dead exist *as* active subjects. Specifically, he challenges us to reflect on two situated death practices as an alternative framework for a critical and creative approach to the innumerable deaths bursting from entangled crises characteristic of the Anthropocene. These are: the fabrication of a counter-cartography in which the public is invited to participate as part of a public search to locate corpses and recover their identities, as well as the reenactment of a forensic ritual that performatively illustrates the right to a proper burial that makes possible the conditions of 'grievability'. This is where I situate what *Necropolis* does best: the articulation of a model for how the living can ethically interact with those who have died unjustly. A model that generates diffracted modes of relations and becomings, which entails the necessity of moving from virulent individualism to a conception of relational subjectivity, no longer contained within an educational paradigm of resilience where the subject is (or should be) apolitical, competitive, and docile, but, instead, vulnerable, in the sense that Michalinos Zembylas (2019) understands it: 'as an openness to an encounter with the other', here with uncanny, marginalised others, 'the point of departure for ethical transformation' (p. 103). *Letting the dead teach*, as the title of this essay suggests, evokes such a relational shift in thinking our relationality with the dead, the latter compelling Zaides to create a difficult but essential artistic experience for learning what needs to be done in the reality of life lived with others, alive and dead.

Necropolis (2017)

The fundamental tool providing context for this work originates in the 1993 UNITED list of refugee deaths, regularly updated in the archives of the non-governmental organisation UNITED for Intercultural Action, which groups together over five hundred anti-racist organisations in forty-eight countries. Since its creation, up to May 2024, the list had recorded 60,620 deaths due to 'the fatal policies of Fortress Europe' and includes basic documentation on each of these cases (see https://unitedagainstrefugeedeaths.eu). In perusing the document, Zaides and his collaborators noted that one could 'hardly ignore the fact that only a very small number of the deceased were mentioned by name, leaving the vast majority of the reported dead without identifying details' (Zaides, 2022, para. 5). *Necropolis* is a collective investigative effort, leveraging the UNITED list as a central database and complemented by active research to locate all remaining burial sites of refugees and undocumented asylum seekers and to identify unreported deaths. This dataset overflows with neglected and harmed bodies, geographies, histories, and displacements, which the team meticulously assembles and converts into an ever-growing computer-generated cartography, laying the groundwork for an architecture of death: a multimedia, choreographic performance combining documentary images, counter-mapping, and stage installation. The artwork has been preceded by extensive preparation and research (Zaides, 2021; Stalpaert et al., 2021) and follows rigorous protocols and procedures, echoing the scientific/artistic/activist work of the Forensic Architecture collective in its quest to hold states accountable for the violence they commit (see Institute of Contemporary Arts, n.d.).

Making Visible the Invisibilised with Counter-mapping

The performance begins with a text projected onto the screen, announcing that *Necropolis* is not a show nor a usual performance to be applauded. Rather, it is the presentation of an ongoing process of archival work. The set consists of an austere, dimly lit room, with a large screen in the background and a high-tech workstation in the foreground. Zaides and fellow researcher Emma Gioia, seated at their computers, turn their backs to the audience. A spatial map is projected onto the screen, displaying Google Earth 3D satellite images of the performance venue. The unhurried, virtual journey begins, unveiling the ever-expanding cartography emerging beneath the European topography and its necropolitical zones—the Mediterranean Sea, the English Channel, and the Balkan Route. It always begins from the geographical position of the host theatre. A first red pin pops up, marking the nearest grave a few kilometres away from Ghent, Belgium, where the theatre is located. Its geographical coordinates appear on the map, followed by a brief description: name, place of birth, and cause of death. This is enough to make you shudder.

The use of Google Earth and Google Maps, originally developed for military purposes and territorial control, is repurposed to challenge their underlying history of racialisation and dehumanisation. This counter-mapping seeks to disrupt dominant power relations by providing sensory and affective understanding of the materiality of social exclusion that kills (socially and physically), and to reveal the destruction of marginalised bodies kept out of the mind and sight of everyday European citizenry. This morbid visual field is enhanced by a disembodied voice that magnifies the voices of the dead, drawn from oblivion, which addresses the audience directly. From the outset, the audience is caught within a reversed situation: They are the ones entering foreign territory to which they do not belong.

> In order to gain the right to live in Necropolis, one has to die in an attempt to enter it. Citizenship is granted posthumously to dismembered, decomposing corpses.
>
> (*Voice-over from* Necropolis)

As the operators slowly zoom in on the exact location of the burial site, Google Earth fades to reveal documentary footage of the physical cemetery. An unknown visitor films their walk from the entrance to the grave.[2] This choreographed ritual is repeated according to a pre-established protocol, performed each time a grave is found. The walk toward the grave serves to illustrate an 'embodied practice' so that the researchers (and the public) can approach the body from an 'eye-level perspective' and pay their respects at the same time (Stalpaert et al., 2021, p. 16). Drawing on Donna Haraway's essay 'Situated Knowledges', Stalpaert et al. (2021) explain how the visit to the cemetery articulates an 'embodied investigation', enabling the localisation of knowledge, creating an interesting tension with Haraway's critique of the 'gaze from above', enacted by the virtual tour through the mapping devices used to locate the graves. The entanglement of these two perspectives, they note, blurs the boundaries between binaries, allowing us to think jointly of death's immateriality and materiality, unsettled past and unsettling present, and moving beyond modern conceptions of the separation between life and death, absence, and presence. A liminal space is created, interweaving 'geographies, mythologies, histories, movements, and anatomies of individuals who lost their lives on the way to Europe' (Stalpaert et al., 2021, p. 16). By setting these tools in motion (cartography, text, film, sound recording), the aesthetic framework of *Necropolis* thus veers away from the reproduction of an ethereal, transcendental universe separated from material bodies and earthly ecologies. Rather, it acknowledges and addresses the relationship between sites of destitution and hegemonic power while materially, historically, and locally situating the dead in order to see reality from a politically informed understanding of the present.

This is precisely what Despret (2021) suggests a work of art can do: it 'fabricates' through 'fabulation' (pp. 6–7) fragments of a world that makes new realities possible—here, the redrawing of a thanatic territory in which the entangled voices of the dead are not reduced to silence but speak from within the abusive laws of racial/nationalist capitalism (Walia, 2021). Not only does the artwork give the dead visibility, it also offers them, at last, a refuge. For Despret (2021), to bring the dead into existence, the question of spatiality is key (more so than temporality, as is the case with Derrida)—she writes: 'wherever the dead are active, places are designated—in a very concrete fashion' (p. 9). The construction of *Necropolis* as an aesthetic and technologically mediated landscape literally 'makes space' for the dead to be 'instaured' (Despret, 2021, p. 9), and, in doing so, defies and resists the logic of erasure and closure that holds them in a supposedly fixed and resolved past. What I find particularly relevant for pedagogical thought is that this act of *instauration* is both relational and ethical, as 'it engages the responsibility of the one doing the *instauring*, to welcoming a request' (Despret, 2021, p. 7). Such request is conveyed (*inter alia*) by the voice-over that addresses the audience directly:

> As you walk out into the street, remember. Every blade of grass pushing its way through the pavement is growing out of their rotten flesh. Living trees are their tombstone. The air that you breathe is their sigh. You inhale, and they are inside of you. You exhale, and the wind blowing through the branches speaks in their voice. Please listen.
>
> (*Voice-over from* Necropolis)

The act of instauration thus reveals how death engenders certain forms of relationality and response-ability, in the sense Haraway (2008) intends—namely, a process of making each other capable of response, which in this case entails, to 'listen' and 'remember'. Zaides's powerful proposition grants a certain level of subjectivity and a new form of existence to the 'dead-other "*othered*"' (Zaides, 2022; his emphasis), which, as Despret points out, does not mean that they have the ability to act *in* the world, but rather that they are conceived as *enablers.* In this sense, through their *instaured* presence within the work, the dead *enable* the audience to think, feel, or perceive otherwise, in a way that grounds the possibility for the living to face up to their response-ability.

In substance, the contemplative itinerary through the redrawn cartography exposes a thousand marked and unmarked graves, implicitly revealing the everyday violence and injustice entangled with the present realities of audiences. For instance, the disclosure of nearby graves surrounding the theatre and unfolding deployment of geographical coordinates produces a disturbing, affective force that directs our attention to the fact that migrant deaths and brutal treatment are happening right next door, not merely in distal places, in the past, or confined to iconic events such as the Lampedusa shipwreck. Moving beyond representations, *Necropolis* productively collapses binaries and, to some extent, subjectivities (compelling one to ponder: Who is now living? Who is dead?) in way that reveals the constitutive ghostly entanglements running through the daily life of the European subject as citizen and rights holder: the many hi/stories of life and death that continue to make contemporary Europe. Hence, regardless of the cultural and political forces that impose oblivion onto the dead, the audience bound to these historical and social effects of loss, exclusion, and death is spatially caught in a 'fabricated' necroworld free of an instituted border politic (be it geographic, epistemic, or ontological). *Necropolis* is built in excess of 'Fortress Europe' and its states of surveillance protected behind the verticality of thick walls erected because of the very fear of their disintegration. In this sense, the dead of

Necropolis are not messengers of hope or despair, nor vectors of guilt, but they confront the audience with the reality of the expendability of their lives and anonymity of their destruction, provoking vivid interrogations on the systemic suppression of the ethical and political significance of unofficially recognised dead bodies and their invisibilisation from white-centred social life. This reality refers to Achille Mbembe's reconceptualisation of sovereign power as the exercise of and control over mortality, termed 'necropolitics' (Mbembe, 2003, 2019). These death-optimising mechanisms, engendered by the traces of events (colonialism, modernity, neoliberal capitalism) that persist in subjecting the *non-citizen* (human and otherwise), enter a death process where bodies are 'let to die' or killed with impunity away from public sight. To counter this disavowal of death, *Necropolis*'s nonhuman voices invite the audience, in non-instrumental yet confrontational ways, to bear witness to the kind of world that is built when some lives are deemed expendable and, even worse, sacrificed for the creation of the physical and ideological boundaries of the sovereign state.

By forging a liminal space where the 'dead-other *othered*' speaks, the artwork's unsettling address opens up new spatio-temporalities, affects, and thought, compelling the audience into the role of witness, if only in the instant of a fragile encounter. At this juncture, an ethic of visibilisation and remembrance emerges because the modes of perception and creation have shifted and the dead are no longer produced by or for nationalist, racial-capitalist narratives of the ruling elite. In this way, as an amplified pedagogical and politically informed space, *Necropolis* has the effect of bringing its viewers outside of a state-based politics of memory and beyond the anonymisation of statistics (Shildrick, 2020). This is how I see the Zaides project's desire to 'make memory *with*' the dead, with a view to activating what Julie Gibson (2021) describes as 'counter-memory', i.e. the 'refusal to forget' (p. 29), which she places at the heart of a 'death ethic' for tackling environmental injustices of the Anthropocene. The 'with' is essential because it allows us to move away from a pedagogical approach that extracts and controls knowledge and unambiguously transmits past stories in the paradoxical hope of rectifying the past, or worse, reburying it (Shildrick, 2020). This, to me, means veering away from tendencies in education that instrumentalise the dead as objects of learning, and heading instead toward relational, fabulist, and open-ended pedagogies, engaging the memory of those who died unjustly as part of our shared world in order to resist forgetfulness and amplify horizons of thought in the here-and-now. This approach evokes what Sharon Todd (2015) sees as the possibility of an educational approach to injustice: one that 'outlines commitment to confronting what is in the present' (p. 54)—an approach which in this case brings existences together, not under an idealised view of a common humanity, but as part of a staged, ethical, 'aesthetic encounter' (after Todd, 2023) between lives that live and die in close proximity to one another but that have never touched.

Therefore, while individuals in the audience are not inherently guilty of violence and harm, *Necropolis* nonetheless grounds the capacity for collective response by creating the conditions in which it can be assumed but nonetheless requires a willingness to receive teachings from the dead, one that involves making oneself vulnerable as the springboard for engendering sites of 'ethical responsiveness' and 'political action' (Zembylas, 2019). Taking my cue from Gert Biesta (2014), 'such teachings often provide insights about ourselves and our ways of doing and being—insights that we were not aware of or rather did not want to be aware of' (p. 96). Yet, when we allow the dead to teach us if we agree to 'welcome the unwelcome', it reveals to us that death is not just an obstacle, but a vector for an affirmative ethic of transformation (Braidotti, 2019). This affirmative orientation is here based on highlighting what needs to be confronted in the present: namely, what is not accounted for, what can/should be resisted, what is possible, and what is left to be done—not as something to be achieved once and for all, but as an ongoing, active task of educational worth.

The Counter-forensic Ritual and Search

How to move beyond representations of these denied deaths, not merely as a haunting but to bring solace to the dead, is a profound pedagogical question that Zaides grapples with. If the slow voyage evokes ghostly traces—the injustice that haunts—the audience encounters the dead, as we have seen so far, in a very material and situated way, thanks to data collection, identification, and an emphasis on place. In the second part of the piece, however, the counter-forensic re-enactment reveals the fault lines of death's (un)representability, and it is here that pedagogical tensions unfold more clearly. The journey through the digital landscape is interrupted rather abruptly by a dramaturgical shift: The performers leave the stage and come back with a forensic table lit by bright red spotlights, on which lie hyper-realistic, gory-looking human body parts. With focused care, they both reorder the artificially made, freshly burned and calcined limbs, torso, and head, while large scanned images of each part are projected onto the screen. This process emblematises performatively a forensic autopsy—'a procedure that a lot of the bodies of migrants who died at Europe's doorstep never have access to' (Zaides, 2022, para. 23). These (dis)assembled, misproportioned human remains seem to metaphorically reflect the dislocation and dispossession of migrant lives and highlight the connections of a singular fate with large-scale networks of transnational migratory flow, part of the new ecology of human/nonhuman mass death in the Anthropocene. Piecing together parts that come from different bodies to form a single body is provocative in that it exposes an inconvenient truth: the degrading and collective reality of death in migration. This materialising of violent death could be read as something compelling, for it addresses unequivocally the invisibility of the exercise of power over the reality of harm and humiliation of invisibilised bodies, placing the corpse not merely within a biological category, but within a product of necro-politically situated matter (Edwards, 2018). By reconstituting a body, the intent of the ethical gesture of performing the counter-forensic ritual is potentially convincing, since it tells a story about the response-ability of the living to address unresolved claims (the right to be identified, named, and grieved). This autopsy of the inautopsiable, this act of counter-forensis invites the audience to collectively experience the transmission of another form of mourning—namely, one that *re-activates*, to use Despret's wording, the beneficial power of ritual, conceived as a form of repair to what has not been done—and performatively restores grievability. Moreover, by staging a ritual, Zaides transposes what many of us would consider a private and personalised event into one that suddenly touches on collective life, transforming the dead of a few into what Despret (2023) calls 'our common dead' (p. 14, my translation). By this, she means that death becomes the vector of a common concern—a shift that the work of art as a public place makes possible.

As a member of the audience, however, I was slightly taken aback by this macabre representation and how two performers lorded over massive body parts representing actual people's suffering and making that on display. Not only does this artistic intervention cast doubts on its ethical intentions, bordering on the subjective imposition of a body of research, it also calls into question its pedagogical contribution, making the work vulnerable to criticism of being more performative than educationally substantive. In contrast to the overwhelming number of dead as they previously appeared on the map, singular in nature, the corpse generalises. Indeed, as it is meant to represent *everyone*, exemplifying all bodies as 'the' dead body, it erases singularity, hides specificities, and instead embodies mass destruction—a sort of materialist haunting that may be the affect that the artistic project intends to convey. What is questionable, from a pedagogical standpoint at least, is the risky proposition of recomposing a body from disparate fake fragments, evoking a rather Frankensteinian imaginary. Given such an imaginary, one wonders what ethico-political orientation the piece might suggest, since it could refer to the prejudicial articulation of the perceived monstrosity that weighs upon the migrant subject who is denied of humanness—the same framework that responds to forms of justification for their rejection and killability. In my view, this entails a significant pedagogical risk, for it relays a sense of essentialisation of the migrant subject, situated in a necropolitical continuum in which they are forever condemned, fabricating a social imaginary solely in relation to his or her tragic fate. In this regard, Despret (2021) aptly suggests that 'a tragic death often reconstructs the life of the victim as if it could only lead to that tragedy: that's all that is remembered, as if everything that had gone before had been turned towards death—so we kill people twice' (p. 26).

From this perspective, 'making memory *with*' the unjustly dead means more than acknowledging their tragic end. While Despret's stance on the dead does not flatten singularity, Derrida's project lends itself to generalisation—two visions that are not mutually exclusive but important to keep in mind when developing pedagogies focused on absent presences, since their 're-presentations' (Despret, 2023, p. 18) are integral to how we apprehend the dead and cultivate a capacity to respond.

Beyond this critique, the re-enactment also metaphorically serves to convey the necessity of the formation of a community of counter-forensic researchers, originally founded by Zaides and his team. From the outset, *Necropolis* offers to all the opportunity to participate in the public and participatory forensic search of thousands of bodies still awaiting identification. By presenting an fixed protocol accessible via an Internet link, the artist offers a potent pathway to audience members (and beyond) to engage in a transformative, ritualised process 'to share a space with them and to confront the fact of their death, not in order to undo what has happened (that cannot be done) but to transform their silence and disappearance into names, stories, and claims' (Keenan, quoted in Zaides, 2021, p. 342). In crossing the boundaries of the performance, this invitation further helps to obscure the real/imaginary divide of seemingly separate worlds (i.e., the living and the dead; the victim and the implicated; the spectator and the watched) through the concretisation of a possible lived encounter with the 'dead-other *othered*' in their actual place of death. The ethical task of recovering an identity in a profound way gives an existence meaning that until now has had no social visibility or significance. Yet, of course, this process points to 'the final impossibility of full justice', as Shildrick (2020) notes, and serves to remind that 'what is necessary is to move beyond the idea that historical wrongs can be brought to light and through a process of censure, shame and recompense be effectively reburied. That cannot be what living well with the dead entails' (p. 176). *Necropolis* follows this line of thought in holding the present open to the non-ending nature of its project that has 'no clear beginning or end since this kind of killing doesn't stop and will probably escalate' (Zaides in Noeth, 2020, p. 52). By the fact that *Necropolis* can never be an 'accomplished' work in the sense of being final and finished, it shows us that an ethical response to the dead is indeed 'accomplishing' something but needs to be constantly renewed (Despret, 2023).

Going beyond the simple act of making visible, this invitation and its protocol call upon the responsiveness of those who are addressed to expand and give a future to *Necropolis* and complete the UNITED list by finding the exact location of a grave, retrieving information on a particular case, and paying due homage. For Despret (2023), while a protocol is 'indeed a political gesture', it is 'above all an artistic gesture … because it is *about* transforming something in the world … through the work and beyond' (p. 13, my translation). In my view, the protocol also becomes a pedagogical tool built into the artwork, insofar as it is implicating those who 'listen' and 'remember', enabling a transformation: doing politics and collectivity differently, on behalf of those who can no longer do so, yet 'starting from them' (Despret, 2023, p. 21, my translation). This pedagogical process widens solicitude for the dead to repair what could not be achieved (identification, burial, ritual) and brings into being, and makes present, a world generally reduced to silence. This gesture of collective intention is one possible response to the difficult task of ethico-political subjectivity formation as itself embedded and implicated in the very things one rejects and wishes to change. Rather than seeing implication as a disinvestment from a situation, it can be seen as genuinely generative, in the sense that it allows something different to emerge (Rothberg, 2019). By realising that subjects are always and necessarily implicated in the power structures that fuel the issue we are attempting to resolve, this allows for the possibility of endless change, but one that presupposes making oneself open and therefore vulnerable to atrocities rather than resilient. As already implied, I understand vulnerability in pedagogical terms not as a projection of oneself and one's affective reactions onto the 'other', but rather as that which enables an encounter with the other and heralds potential ethical transformation. The invitation to research puts this pedagogical orientation into practice.

Despite the tensions this performance piece is bound to have (such as in the issues of representation, appropriateness, and the aesthetisation of death), the passage it opens is an encouragement *not* to forget and to honour the memory owed to the dead by ensuring 'the right of the disappeared to be sought for and identified, and to uphold the rights of their families during such search processes' (Zaides, 2022, para. 2). In this sense, the artwork is not a strategy of consolation, or a way of reducing the dead to a didactic unit that would bolster individual resilience in the face of harm done to others (more easily denied than when inflicted on oneself) and simply allow us to move on. Rather, it tries to shift the educational terrain toward an openness to the difficult and ongoing task of questioning what it means to live lives in the shadow of mass death and to seriously consider the creation of ethico-political means of honouring and remembering the dead, including (and this is an important gesture) preventive recognition of the destructive consequences of violence on those directly affected. As the voice-over pleads:

> They are the forgotten ones. They are the ones that you cannot forget. The ones whose dwelling is never built, whose arrival is forever postponed—they are the ones who are silently speaking to you now. They are speaking from a locus of absence, of erasure, of justice never delivered. They speak in memory.
>
> *(Voice-over from* Necropolis*)*

Here it becomes manifest that the thoughtful arrangement of the piece to forge a *common* 'counter-memory' is of pedagogical importance, not only because it brings the unspoken to the surface (in Derridean mode), but also because, in line with Despret's methodology, it enables us to understand what needs to be done in practice for every other singular, vulnerable individual and community, (human and others). Following Despret, this is precisely what the artwork allows for: The violence of death does not cancel out the possibility of the metamorphosis that follows. This non-prescriptive, unpredictable metamorphosis becomes the subject of creative work. Ultimately, what she conceptually helps us to see, and which Zaides materialises, is the potentiality of the artwork to form more just and inclusive modes of relationality and becoming that take the dead out of their spectral dimension and make them more common, literally, by giving them a 'place' and making them '*in* common'. That itself is a pedagogy, because it is literally teaching us to live with the dead as opposed to casting it as a giant, haunting unknown—which, in these times, often creates affective modes of powerlessness and disengagement.

Conclusion

This chapter aimed to illustrate that any pedagogy oriented toward a 'justice-to-come' (Carstens, 2021), wishing to contribute to a transformative paradigm for and in a post-human world, cannot ignore death and must attend to those whose lives have been unjustly taken away. In staging an encounter with the dead, *Necropolis*'s pedagogy opens up non-normative possibilities for thinking, perceiving, and relating with death in a way that is more expansive and ethically in tune with the many harrowing realities of entangled crises that are so difficult to grasp but in which we nonetheless live. As a teaching deeply rooted in the artwork, the ethical necessity of re-subjectifying the dead is manifest (Despret, 2021) and parallels the idea that death solicits response-ability, with death itself becoming an exhortation to remembrance. Thinking with this piece of art thus makes apparent that learning from/with the dead does not merely account for the destruction of bodies, but also for their *construction* as part of fabricating an ethico-political and an educational space constitutive of the present. Without erasing contested pasts and ongoing systemic structures of violence, the artwork extends possibilities to bring the past into the present, giving it a 'chance' to transform the future (Despret, 2023, p. 16).

Necropolis, not purely artistic nor self-referential, in a way replaces what European states and policies neglect, thereby enacting what is not being done but can be. In so doing, it also paradoxically shows the pedagogical limits of art, in that ethical interactions between the living and the dead cannot simply be immortalised in the artwork (hence the warning that it is not a standard performance). Nevertheless, it also opens up avenues for overcoming this limitation. The practices and tools included in the model created by Zaides (counter-forensics, counter-cartography, and other small death rites) are examples of this. That is why I think this type of ethical model deserve to be considered in pedagogical thinking concerned with posthuman, hauntological, and decolonial issues, because without them, the temptation to label the dead too easily as lost or forsaken causes will persist (Gibson, 2021). Moreover, while these are not easy relations, not least because they run counter to dominant forms of exchange based on subject-object dynamics that prevent encounters with the 'other-*othered*', they also disrupt the multiple boundaries of time, space, subjectivity, and memory, inviting a certain vulnerability and openness in the face of death. By seeking to actively mend broken connections through the process of creating 'memory *with*', Zaides shows the creative and critical possibilities (and limitations) of constructing two worlds in dialogue, so that we, as living people holding unearned privileges of political citizenship, can see ourselves entangled with them rather than as observers and alienated from these necro-conditions and lives. For me, this is educationally encouraging, for it leads us to consider the power of art to fabricate/fabulate galvanising new ways of relating to premature and unjust death, in a more affirmative manner to denial or contempt, and to begin to create new imaginaries of death and different worlds that establish the conditions for transformative action. Here, *Necropolis* has opened a passage to draw attention to the significance of migrants lives against extermination, not in disavowal of the negative, but precisely at the heart of it, in a process of potential, for the creation of more just and liveable futures.

Notes

1 Although the artwork focuses on migration death specifically, its content and pedagogical contribution remain relevant to the present publication. For the fundamental reason that transnational migration will only intensify and accelerate due to climatic disruption caused by global dependence on fossil fuels linked to capitalist models of extractive and exploitative production. Additionally, the rise of fascist politics and racialised nationalism continues to consolidate systemic and structural injustices of the migration regime, based on the operation of Eurocentric matrix of power linked to the project of modernity and its colonial history of violence, separation and domination (Mignolo, 2011). The colonial/modern nexus is considered to be at the heart of the Anthropocene (Davis and Todd, 2017), the epoch turned marker of all eco-social injustices, where, among other problems, vast migratory movements entangle humans and more-than-humans in mortal assemblages as climate change and ecological devastation fundamentally redraw the map of habitable places, endangering the planet's most vulnerable communities.

2 Here is an example of such a short film projected during the performance: https://vimeo.com/730292404

References

Banerjee, B., Misrahi-Barak, J., and Lacroix, T. (2023). Introduction: Thanatic Ethics: The Circulation of Bodies in Migratory Spaces, *Interventions*, 26(1), pp. 1–20. https://doi.org/10.1080/1369801X.2023.2190917.

Biesta, G. J. J. (2014). *The Beautiful Risk of Education*. Boulder, CO: Paradigm Publishers.

Braidotti, R. (2019). Affirmative Ethics and Generative Life, *Deleuze and Guattari Studies*, 13(4), pp. 463–481. https://doi.org/10.3366/dlgs.2019.0373.

Butler, J. (2004). *Precarious Life: The Powers of Mourning and Violence*. London: Verso.

Carstens, D. (2021). A Posthuman Hauntology for the Anthropocene. The Spectral and Higher Education, in V. Bozalek, M. Zembylas, S. Motala, and D. Hölscher (eds), *Higher Education Hauntologies Living with Ghosts for a Justice-to-Come.* London: Routledge, pp. 120–134.

Davis, H. and Todd, Z. (2017). On the Importance of a Date, or, Decolonizing the Anthropocene, *ACME: An International Journal for Critical Geographies*, 16(4), pp. 761–780.

Derrida, J. (1994). *Specters of Marx.* Translated by P. Kamuf. New York: Routledge.

Despret, V. (2021). *Our Grateful Dead: Stories of Those Left Behind.* Translated by S. Muecke. Minneapolis, MN: University of Minnesota Press.

Despret, V. (2023). *Les Morts à l'œuvre.* Paris: Éditions La Découverte.

Edwards, E. E. (2018). *The Modernist Corpse.* Minneapolis, MN: University of Minnesota Press.

Gallien, C. (2018). 'Refugee Literature': What Postcolonial Theory Has to Say, *Journal of Postcolonial Writing*, 54(6), pp. 721–726. https://doi.org/10.1080/17449855.2018.1555206.

Gibson, J. (2021). Climate Justice for the Dead and the Dying: When Past-Oriented Environmentalism Isn't Enough, *Environmental Philosophy*, 18(1), pp. 5–39.

Haraway, D. (2008). *When Species Meet.* Minneapolis: University of Minnesota Press.

Institute of Contemporary Arts. (n.d.). Forensic Aesthetics, www.ica.art/forensic-aesthetics (accessed 10 September 2024).

jagodzinski, j. (2015). Affirmations and Limitations of Rancière's Aesthetics: Questions for art and its education in the Anthropocene, in N. Snaza and J. A. Weaver (eds), *Posthumanism and Educational Research.* New York: Routledge, pp. 121–133.

Lustgarten, A. (2015). *Lampedusa.* London: Bloomsbury.

Lykke, N. (2022). *Vibrant Death.* London: Bloomsbury.

Mbembe, A. (2003). Necropolitics, *Public Culture*, 15(1), pp. 11–40.

Mbembe, A. (2019). *Necropolitics.* Durham, NC: Duke University Press.

Mignolo, W. D. (2011). *The Darker Side of Western Modernity.* Durham, NC: Duke University Press.

Noeth, S. (2020). Interview with Arkadi Zaides: 'Bodies as Evidence. Arkadi Zaides on his Research, People on the Move, and Brutally Closed Borders', *Tia Magazine*, pp. 50–53.

Robertson, J. and Murphy, J. (2017). *The Jungle.* Theatrical play, Young Vic, London, 7 December 2017–9 January 2018.

Rothberg, M. (2019). *The Implicated Subject: Beyond Victims and Perpetrators.* Redwood City, CA: Stanford University Press.

Shildrick, M. (2020). Queering the Social Imaginaries of the Dead, *Australian Feminist Studies, 35*(104), 170–185. https://doi.org/10.1080/08164649.2020.1791690.

Stalpaert, C., Zaides, A., Lussault, M., Rekacewicz, P., Atelier Cartographique, and Dobricic, I. (2021). Tentacular Thinking in Storied Places: A Deep-Mapping of an Art-Science-Activist Worlding in *Necropolis* (2017–Onward), *Global Performance Studies*, 4 (2). https://doi.org/10.33303/gpsv4n2a5.

Todd, S. (2015). Creating Transformative Spaces in Education: Facing Humanity, Facing Violence, *Philosophical Inquiry in Education*, 23(1), pp. 53–61.

Todd, S. (2023). *The Touch of the Present.* Albany, NY: SUNY Press.

Walia, H. (2021). *Border and Rule: Global Migration, Capitalism and the Rise of Fascist Nationalism.* Chicago, IL: Haymarket Books.

Weiwei, A. (2017). *Law of the Journey.* Sculpture, National Gallery, Prague, 17 March.

Zaides, A. (2021). Necropolis: Walking through a List of Deaths, in D. Agostinho, S. Gade, N. Bonde Thylstrup, and K. Veel (eds), *(W)archives: Archival Imaginaries, War, and Contemporary Art*. London: Sternberg Press, pp. 337–364.

Zaides, A. (2022). Necropolis: Counter Forensic Practices for Mourning the 'Othered' Dead, www.rektoverso.be/artikel/necropolis-counter-forensic-practices-for-mourning-the-othered-dead (accessed 16 July 2024).

Zembylas, M. (2019). The Ethics and Politics of Precarity: Risks and Productive Possibilities of a Critical Pedagogy for Precarity, *Studies in Philosophy and Education*, 38(5), pp. 95–111. https://doi.org/10.1007/s11217-018-9625-4.

8

ENABLING ART TO BECOME A PEDAGOGICAL AND THERAPEUTIC TOOL IN REFUGEE CAMP CLASSROOMS

Paul O'Keeffe

Introduction

The United Nations High Commission for Refugees (UNHCR) estimates that there are close to 110 million people forcibly displaced from their homes around the world (UNHCR, 2022). Of these, some 4 million people (or 10% of the global refugee population) have sought sanctuary in just four countries of East Africa—Uganda, Kenya, Sudan, and Ethiopia (Oxfam, 2023). These countries, for the most part, while already heavily burdened with their own internal social, political, and economic problems, operate warehousing-like approaches that contain refugees in camps far from domestic population centres, amenities, and opportunities. In fact, there are around 300 such camps in the wider East Africa region, usually situated in volatile border areas, where millions of asylum seekers and refugees are confined by the strictures and structures of a forced migration management system that offers little more than a place to wait out turmoil, destitution and even death in their home countries (Camarena, 2023).

Adding to the turmoil caused by political and social strife in the region, climate change is having a dramatic effect of peoples' lives in East Africa. After 5 consecutive seasons of drought, torrential downpours in the 2023 rainy season have resulted in more devastation to food production, and in turn the region's human security (WFP, 2023). Consequently, refugee camps in the region, normally reserved for people fleeing persecution, are currently seeing their populations swell, as desperately hungry people seek sanctuary from the ravages of hunger visited on their communities (UNU, 2023). While not recognised under international law as refugees, people fleeing the worst ravages of climate change have little choice in the region than to seek sanctuary within the confines of the international forced migration management

DOI: 10.4324/9781003456636-9

system. Currently in 2024, 2.3 million people have been internally displaced in Somalia, Kenya, and Ethiopia due to drought and 264 million more have crossed into neighbouring countries (Credendo, 2024). With the outlook for climate induced shocks in the region to worsen and further affect migration patterns in the future, it is inevitable that many more millions of people will come to rely on the safety, protection and support offered in refugee camps (Ehui and Kumari Rigaud, 2022).

While most refugees in the world live outside of refugee camps and generally have access to the goods and services that their non-refugee counterparts enjoy, for the 6.6 million refugees globally who live inside camps, life is very different (UNHCR, 2023). These camps are known as 'contained camps' as they supposedly contain all the goods and services that refugees need within the confines of the camp structure and entry and exit to these camps is strictly controlled. For many theorists and no doubt refugees alike, refugee camps are 'camps of containment' that, reflecting their contradictory nature, provide 'care' but at the same time 'contain' refugees—in the sense that they are contained in a space that they are unable to freely leave (McConnachie, 2016). At a time when more and more people are fleeing to such places to escape persecution and climate induced crises, this contradictory nature may further reflect the sense that the camps are the only places in the region to contain resources that will keep them alive, while at the same time contain them behind camp barriers so that they can no longer live as they please.

Education is a key service provided in contained camps. While often delivered through the national hosting country's education system, schooling in contained camps is a different experience than that experienced elsewhere in the host country. Exacerbated by a 'sense of emergency' that is ever present in refugee contexts, day-to-day services and their administration often follow the 'exception' and not the rule. Furthermore, schools in contained camps tend to be oversubscribed, under resourced, with their students experiencing an array of educational barriers, including deep psychological traumas and negative consequence of multiple interruptions to their educational pathways (Hampel, 2022).

In this chapter I explore the educational environment of refugee camps and the psycho-social implications that education can have on children in developing and crisis contexts. Having spent over a decade working on refugee and education development projects, both as an academic and development practitioner, in Europe, Asia, the Middle East and East Africa, I draw on my experience of refugee education and the development discourse which surrounds refugees living in contained camps, with the aim of proposing some pedagogical reforms for the betterment of the lives of child refugees. In doing so, I put forward the idea that the development and incorporation of art therapy approaches in curriculum building and pedagogical practices in refugee camp schools should become one solution that could benefit an increasing number of children in an uncertain future.

The Case of Kakuma Refugee Camp

Kakuma refugee camp in Northwest Kenya, is typical of contained refugee camps in the region. Formed in the early 1990s, the camp was initially designed as a temporary sanctuary for Sudanese youth fleeing civil war, and now, along with the adjacent Kalobeyei integrated settlement, is home to close to 250,000 people from around 24 different ethnic communities (UNHCR, 2019). Conditions in Kakuma can be said to resemble that of large shanty town-like cities in the region. Extreme temperatures, dust roads, makeshift homes, poor if any sanitation, overcrowding, lack of health care and other services, disease, and hunger characterise life in the camp (Carron, 2019).

With the average time refugees spend living in Kakuma exceeding 19 years, the temporary nature of its tarpaulin covered homes and makeshift infrastructure is at odds with the reality faced by its residents (Fairfield University, 2023). When asylum seekers initially arrive in the camp, they are issued some space on which to build a house and some wooden poles and corrugated iron or tarpaulin to use for roofing for self-made mud huts under the general assumption that they will return home or, for the lucky few, be relocated abroad in a short time. In addition, they are usually given a mosquito net, and monthly food rations which amount to roughly 70% of what they need to survive and little else (Refugees International, 2017). Due to legal limitations on refugees working in Kenya and the difficulties in trying to produce their own food with agriculture (Kakuma is located in a semi-arid region where temperatures regularly exceed 40 degrees Celsius), people who arrive in the camp, find themselves in a situation where they cannot work to support themselves, cannot leave the camp easily and face hunger and desperation while they wait a very long time for their prospects to improve (WFP USA, 2024).

Consequently, and rather inevitably, a shadow economy has grown up in and around the camp which the refugees participate in to substitute their survival on the handouts that they receive from the international humanitarian infrastructure which manages the camp. To make ends meet, people eek out livings selling everything from charcoal made from local shrubs to food vouchers given to them in lieu of monthly rations. Kakuma, for all its problems, is a living and breathing city where hundreds of thousands of people go about their business seeking opportunities and better futures, wherever they can find a chink of hope (NRC, 2018).

Education in a Kakuma, and most other refugee camps, mirrors people's experiences of accessing food and other necessities—it is available at a minimum and just about enough to get by while one waits for something better to come along. Roughly half of Kakuma's population are children and of these 100,000 + children, only 36% regularly attend primary school (Mendenhall, 2018). While recognised as a basic human right (UNHCR, 2021) education, and access to it, is extremely difficult for most in Kakuma and, as in other refugee camps in developing and crisis contexts, rarely extends beyond basic primary level. Figures in the international development literature put attendance rates at 68% for Kakuma's 26 primary school children (compared with 91% worldwide) and 34% for its 5 secondary schools (compared to 84% worldwide) (UNHCR, 2021). As low as these figures may sound, for those who have seen the insides of Kakuma's classrooms (and other refugee camps in East Africa and beyond), they are wildly optimistic. Lack of resources, the use of corporal punishment in severely overcrowded classrooms combined with inadequately trained teachers and a whole host of other barriers coalesce to result in accessing quality education in Kakuma being an almost impossible task (Crea et al., 2022).

Teaching and learning in such difficult environments as refugee camps attracts a lot of attention from pedagogy and development researchers and scholars the world over. Due to education being a fundamental right and its championed ability to raise people out of poverty, it also receives a lot of attention on the national and international governance stage. Between global compacts and national development plans, billions of Euros are pledged each year to development education by donors and politicians eager to do (and be seen doing) the right thing (Their World, 2014). Despite all these efforts, in places like Kakuma with hundreds of students cramming classrooms and all the associated barriers to education that they face, one would be forgiven for thinking little progress is being made. Compounding this issue is the international forced migration management system's fixation with datafication (Madianou, 2019) and a resulting 'bums on seats' approach that sacrifices quality education in the name of quantity (Finn Church Aid, 2023).

In Kakuma the education system follows the Kenyan curriculum, which is currently undergoing reforms of its '8-4-4' system. This system accounts for eight years of basic (primary) education after which students sit the state Kenya Certificate of Primary Education (KCPE) exams in Kiswahili, English, mathematics, science and agriculture, and social studies. Following this, if a student attends secondary school for 4 years, they can sit the Kenya Certificate of Secondary Education (KCSE) exam in mathematics, English, Kiswahili and a range or other non-compulsory subjects which depend on the offerings of an individual school. Reforms are underway to transform the entire education system to a new format of '2-6-3-3-3', consisting of two years pre-primary, six years of primary (three years lower and three years upper primary), six years of secondary (three years junior and three years senior) and three years of university education (Mwugusi et al., 2022). Reflecting its geographical remoteness, and perhaps its lack of importance on the national educational agenda, education in Kakuma still follows the 8-4-4 system with a rigid prescriptive curriculum, that relies on summative assessment and competition amongst students who rarely have any opportunity to learn beyond repetition of their stressed-out teachers (Mwugusi et al., 2022).

Educational Barriers or Educational Opportunities?

Going beyond the obvious barriers that overcrowded classrooms present to effective pedagogy (the student teacher ratio is over one hundred students per one teacher), the language of instruction used in refugee camp schools (English in the case of Kenya, or Kiswahili in instances where students come from multi-linguistic backgrounds) doesn't always reflect the language needs of refugee students. In underfunded and over subscribed classrooms, students who don't speak the language fluently start their schooling at a disadvantage point that is difficult to overcome. Furthermore, due to the very nature of the experiences that have led to them to become asylum seekers in the first place, children, and young people in refugee camps invariable suffer from a host of psycho-social traumas that negatively impact on their educational attainment (O'Keeffe, 2023). According to the World Health Organization, refugee children are particularly vulnerable to mental health stressors as the context of displacement affects all parts of their lives and happens at a crucial time in their physical, emotional, social, and cognitive development (WHO, 2022). Uprooted from their homes, placed in strange and challenging contexts, and often surrounded by deeply traumatised family members, schooling or more precisely the promise that education can lead to a better life is often the only bit of hope refugee children have (Kubwalo, 2015).

As it stands, curriculum in refugee camps in developing countries are paired back, limited, and limiting. Rarely are students taught anything beyond the basics and, unlike efforts to help heal traumatised refugee children and encourage better integration (as is evidenced within education systems in more prosperous countries), children confined in classrooms in places like Kakuma, while heavily invested in a promise that education leads to better lives, rarely benefit from all that education has to offer (O'Keeffe, 2023). This unfortunate assessment doesn't have to be the case and things can change for the better if there is a will to do so.

Despite all the problems associated with formal education in refugee camps, schools nevertheless offer some semblance of normality for refugee children with troubled lives. Providing security and routine, after uncertainty and fear, schools should be safe havens that play a vital role in helping children adapt to their new lives (UNHCR, n.d.). Furthermore, education also offers an opportunity to help refugee children heal from some of the many traumas that they have faced in their lives. Such a function of education is already well established in wealthier countries, where the psycho-social benefits of schools to help refugee children overcome trauma is well established. Programmes such as the European Union's REFUGE ED, which encourage schools to combine mental health and education best practices (European Commission, 2023) and the USA's federal Head Start programme, which enables refugee children access early childhood education programmes at the state level (Hoagland, 2022), are just two examples of the acknowledgement of, and importance placed on, the role that education can play in healing traumatised refugee children and helping them adapt to their new lives in exile. Unfortunately, this is not the case for most developing countries and especially not for education in contained refugee camps such as Kakuma. With rote learning of basic curricula *de rigueur*, there is little space afforded to the properties and promises of education in such places, beyond preparing students for success in inflexible summative assessments at the end of their time in formal schooling.

Improving Education, Despite the Constraints

At this point, we can turn our attention to how the situation can be improved in refugee camps, so that education helps children and young people deal more effectively with the traumas that they face in the young lives. Answering this is by no means a simple matter and one that requires an understanding of the specific educational needs of refugee children and the idiosyncrasies of life in refugee contexts. Moreover, improving education in refugee camps so that it meets the needs of deeply traumatised children and helps them to better cope with the challenges they face, needs to start at a point of recognition of education's psycho-social and therapeutic abilities (de Wal Pastoor, 2015). With this recognition naturally follows the impetus to improve teacher training in education psychology and therapeutically enhanced curriculum development. While this is all well and good and much needed, the reality is that most teachers in refugee camps in developing countries have no formal training and little prospects of receiving any. For instance, 85% of teachers in Kakuma's teachers have no formal qualifications and 46% have never received any form of training (Gichuhi, 2022).

Compounding the lack of psychological support training teachers working in contained refugee camps normally have, are the almost ubiquitous social taboos against acknowledging and treating mental health issues in such contexts (Golden, 2017). A study carried out by Kisombe in 2020 on the psychological health of children in Kakuma found that most children in the camp were deeply traumatised and suffered from depression, irritability, lack of concentration and withdrawal and that these maladies were seldom acknowledged and largely ignored (Kisombe, 2020). While the research found that counselling was most accessible for children in the camp via their schools and recommended that counselling become part of the curriculum, it also found that teachers did not have adequate training on how to interact with their students and thus should receive training on how to do so if things are to improve (Kisombe, 2020).

The whole scale reform of education policies and practices in refugee camps, while daunting and probably not a realistic option for most camp administrations in developing countries, at least not in the shorter term, does need to be addressed if the global community is serious about our commitment to education justice in the longer term. To get to such a place will take time, but with acknowledgement of the importance of education in various global compacts, sustainable development goals and other commitments to global development and social justice, it at least appears as though the world is going in that direction, for the moment. In the more practical and immediate term, various humanitarian and development organisations have recognised and begun to fill the psycho-social space in refugee camps all over the world (UNHCR, 2023). Taking the lead has been a cacophony of education and mental health charities operating in refugee contexts in developed countries. These organisations have traditionally operated on an extra-curricular basis, offering support to refugee children, and other displaced people outside of normal school hours. This has begun to change dramatically in recent years as social taboos surrounding mental health abate and a greater understanding of the psycho-social, emotional, and educational needs of displaced children are mainstreamed into curricula (WHO, 2022).

Art Therapy as a Psychological Intervention and Pedagogical Tool

The adoption of art therapy as a psychological intervention for displaced children with post-traumatic stress disorder in western countries has gained increasing attention in recent years. When employed in non-threatening educational settings, it has been found by numerous studies to be useful in facilitating the expression of feelings connected with trauma, regulating negative emotions, and providing a sense of relief for refugee children (Annous et al., 2022; Kalmanowitz and Ho, 2016; Akthar and Lovell, 2019; Wahlbeck et al., 2020). This recognition and acceptance of art therapy as a psychosocial support mechanism has paved the way for increasing calls from policy makers and reception and integration practitioners to mainstream the institution of art therapy in refugee receiving centres and schools all over the western world (European Commission, 2022). The European Union, in particular has led the way in integrating art therapy into curricula in its member countries through projects such as the European Union's Schools Education Gateway webinar series and the European Toolkit for Schools which provide learning opportunities and resources for teachers to engage with and enable their refugee students to better cope with the traumas they have faced and their realities of their lives in exile (Schools Education Gateway, 2022).

Integrating art therapy as a mainstream part of the curriculum for those who work with refugee children, is an altogether easier practice for resource rich countries whose teachers receive many years of professional training, both before and during their careers and have relatively easy access to resources. The reality in poorer countries, and particularly in places such as Kakuma, is far from conducive to such approaches. Notwithstanding the limitations on providing adequate art therapy training for teachers who work in refugee camps and the resources that they would need to enable art as a therapeutic form of expression in their overcrowded classes, there are steps that could be taken to incorporate it into pedagogy in refugee camp classrooms. A simple initial approach could take the form of education policy makers first acknowledging the specific mental health needs of refugee students in their education systems and then the promotion of adequate training for all those who work with refugee students. Such training could focus on some of the basic tenets of art therapy such as its universalism, its ability to process and express emotions and feelings and its ability to negate some of the ill effects of traumatic events and to encourage personal growth (Rowe et al., 2016).

Innovations in pedagogy do not necessarily have to involve a seismic shift in the classroom. A simple approach such as incorporating a new teaching point or new materials into an existing curriculum can often yield more positive results than demanding a complete change in approach (Prentice, 2022). Refugee schools in places like Kakuma, already have social studies on their curricula, which could be a natural home for the incorporation of art into its pedagogy. However, to do so would require a considerable amount of investment into resources and a dedication to teachers. This would ultimately amount to a seismic shift in the emphasis camp administrators and host countries place on the importance of education in refugee children's lives.

Conclusion

While education in refugee camps faces many obstacles as it endeavours to improve the lives of those it purports to help, there is a long road ahead before it can realise its full potential to transform our world for the better. In the meantime, there are many positive examples of education's power to heal and grow that can inspire change within the confines of contained refugee camps. The embrace of art into curricula as a vehicle to express feelings and emotions and enable growth in refugee children could become one such example, if it is given the chance it deserves.

As the interconnected global challenges of conflict, displacement, poverty, and climate change increasingly wreak havoc on the world's poorest people, it is becoming harder for the rest of us turn away. In particular, our apathy towards displacement and climate change can no longer be sustained as its effects are felt more and more in our comfortable lives in the developed world. Solutions to the many problems the world faces are hard to come by, but a good place to start isa collective global response towards the needs of children who are most affected by the problems. This chapter has attempted to illuminate one way in which we can do this, but there are many other approaches that we need to consider as well. Notwithstanding a lack of resources, education can transform the lives the world's most needy, as long as there is the will to do so.

References

Akthar, Z., and Lovell, A. (2019). Art Therapy with Refugee Children: A Qualitative Study Explored through the Lens of Art Therapists and Their Experiences. *International Journal of Art Therapy* 24, 139–148. doi:10.1080/17454832.2018.1533571.

Annous, N., Al-Hroub, A. and El-Zein, F. (2022). A Systematic Review of Empirical Evidence on Art Therapy with Traumatized Refugee Children and Youth. www.frontiersin.org/articles/10.3389/fpsyg.2022.811515/full.

Camarena, K. R. (2023). Most East African Refugees Are Hosted Close to Borders—it's a Deliberate War Strategy. 14 May.https://theconversation.com/most-east-african-refugees-are-hosted-close-to-borders-its-a-deliberate-war-strategy-200861.

Carron, D. (2019). The Study and Teaching of Human Rights in Refugee Camps: Learning From Experience. www.europenowjournal.org/2019/03/04/the-study-and-teaching-of-human-rights-in-refugee-camps-learning-from-experience/.

Crea, T., *et al.* (2022). Inclusive Education in a Refugee Camp for Children with Disabilities: How Are School Setting and Children's Behavioral Functioning Related? *Conflict and Health* 16, 53. https://doi.org/10.1186/s13031-022-00486-6.

Credendo. (2024). East Africa: A Climate Crisis That Is Only Expected to Worsen. https://credendo.com/en/knowledge-hub/east-africa-climate-crisis-only-expected-worsen#:~:text=As%20more%20and%20more%20people.

de Wal Pastoor, L. (2015). The Mediational Role of Schools in Supporting Psychosocial Transitions among Unaccompanied Young Refugees upon Resettlement in Norway. *International Journal of Educational Development* 41, 245–254. https://doi.org/10.1016/j.ijedudev.2014.10.009.

Ehui, S. K. and Kumari Rigaud, K. (2022). Climate Migration—Deepening Our Solutions. www.brookings.edu/articles/climate-migration-deepening-our-solutions/.

European Commission. (2022). EU and WHO Call for Support for the Mental Well-being of Forcibly Displaced People through Art and Culture. https://culture.ec.europa.eu/news/eu-and-who-call-for-support-for-the-mental-well-being-of-forcibly-displaced-people-through-art-and-culture.

European Commission. (2023). Combining Mental Health and Education Best Practices Helps Migrant Children Find Their Feet. https://projects.research-and-innovation.ec.europa.eu/en/projects/success-stories/all/combining-mental-health-and-education-best-practices-helps-migrant-children-find-their-feet.

Fairfield University. (2023). Honors Students Bring Kakuma Refugee Experience to Middle Schoolers. www.fairfield.edu/news/archive/2023/march/honors-students-present-walk-in-their-shoes-refugee-experience.html.

Finn Church Aid. (2023). Quality vs Quantity: The Challenges of Providing Education to Refugees in Kenya. www.kirkonulkomaanapu.fi/en/latest-news/articles/challenges-of-providing-quality-education-to-refugees-in-kenya/.

Gichuhi, L. (2022). The Challenges of Training Female Teachers in Refugee Camps. https://inee.org/blog/challenges-training-female-teachers-refugee-camps#:~:text=Most%20refugee%20teachers%20have%20less.

Golden, S. (2017). Assessing Mental Health in Humanitarian Emergencies: A Representative Survey by the Center for Victims of Torture in Kalobeyei, Kenya. www.cvt.org/sites/default/files/FINAL.Kakuma%20CVT%202016%20Survey%20Summary.10%205%202017.pdf.

Hampel, J. (2022). A Lost Generation? The International Failure of Refugee Education. www.theperspective.se/2022/04/15/migration/a-lost-generation-the-international-failure-of-refugee-education/.

Hoagland, I. (2022). Early Childhood Education for Refugees in the United States. https://blogs.gwu.edu/gsehd-real/2022/11/14/early-childhood-education-for-refugees-in-the-united-states/.

Kalmanowitz, D., and Ho, R. (2016). Out of Our Mind: Art Therapy and Mindfulness with Refugees, Political Violence and Trauma. *Arts Psychotherapy* 49, 57–65. doi:10.1016/j.aip.2016.05.012.

Kisombe, E. (2020). Interrogating Refugee Rights to Health: A Case of Psychological Health of Refugee Children in Kakuma Refugee Camp, Kenya. http://erepository.uonbi.ac.ke/bitstream/handle/11295/154011/Kisombe_Interogating%20Refugee%20Rights%20to%20Health-%20a%20Case%20of%20Psychological%20Health%20of%20Refugee%20Children%20in%20Kakuma%20Refugee%20Camp%2c%20Kenya..pdf?sequence=1&isAllowed=y.

Kubwalo, K. (2015). In a Refugee Camp, Education Is the Only Hope. https://reliefweb.int/report/jordan/refugee-camp-education-only-hope.

Madianou, M. (2019). Technocolonialism: Digital Innovation and Data Practices in the Humanitarian Response to Refugee Crises. *Social Media + Society* 5(3). https://doi.org/10.1177/2056305119863146.

McConnachie, K. (2016). Camps of Containment: A Genealogy of the Refugee Camp. *Humanity*, 7(3), 397–412. https://doi.org/10.1353/hum.2016.0022.

Mendenhall, M. (2018). Teachers for Teachers: Advocating for Stronger Programs and Policies for And with Refugee Teachers in Kakuma Refugee Camp, Kenya. *Studies in Social Justice* 12(2), 356–363.

Mwugusi, G., Mpitabavuma, V. and Arok, M. (2022). INEE Guidance Note on Teacher Wellbeing in Emergency Settings Domain 3—Teaching and Learning Kakuma Refugee Camp Contextualization. https://inee.org/sites/default/files/resources/TWB%20GN%20Contextualized%20Domain%203%20-%20Kenya%20ENG.pdf.

NRC. (2018). Supporting Kakuma's Refugee Traders The Importance of Business Documentation in an Informal Economy. https://hrp.law.harvard.edu/wp-content/uploads/2018/09/Business-Permit-Briefing-IHRC-NRC-1.pdf.

O'Keeffe, P. (2023). Art as a Passport for Learning and Healing in Refugee Camps. www.europenowjournal.org/2023/11/20/art-as-passport-for-learning-and-healing-in-refugee-camps/#:~:text=Across%20Europe%2C%20the%20response%20by.

Oxfam. (2023). East Africa Hosts 10% of Global Refugees While Ranking among 20 Least Developed Countries. www.oxfam.org/en/press-releases/east-africa-hosts-10-global-refugees-while-ranking-among-20-least-developed.

Prentice, C. M. (2022). Educators' Positive Practices with Refugee Pupils at Two Schools in England. *British Educational Research Journal* [preprint]. https://doi.org/10.1002/berj.3818.

Refugees International. (2017). UN Food Distributions Are Critical for Somali Refugees in Kenya. www.refugeesinternational.org/un-food-distributions-are-critical-for-somali-refugees-in-kenya/#:~:text=Overall%2C%20refugees%20living%20in%20Dadaab.

Rowe, C. *et al.* (2016). Evaluating Art Therapy to Heal the Effects of Trauma Among Refugee Youth. *Health Promotion Practice* 18(1), 26–33. https://doi.org/10.1177/1524839915626413.

Schools Education Gateway. (2022). Supporting the Mental Health and Well-Being of Refugee Pupils Through Connection and Continuity. www.schooleducationgateway.eu/en/pub/latest/news/mental-health-refugee-pupils.htm.

Their World. (2014). Record $28.5bn Pledged for Education by Developing Countries and Donors. https://theirworld.org/news/record-28-5bn-pledged-for-education-by-developing-countries-and-donors/.

UNHCR. (2019). Kakuma Refugee Camp and Kalobeyei Integrated Settlement. www.unhcr.org/ke/kakuma-refugee-camp.
UNHCR. (2021). Educational Report 2021: Staying the Course-the Challenges Facing Refugee Education. www.unhcr.org/612f85d64/unhcr-education-report-2021-staying-course-challenges-facing-refugee-education.
UNHCR. (2022). Refugee Statistics. www.unrefugees.org/refugee-facts/statistics/.
UNHCR. (2023). Refugees Need Better Mental Health Support amid Rising Displacement. www.unhcr.org/ie/news/news-releases/refugees-need-better-mental-health-support-amid-rising-displacement.
UNHCR. (n.d.). Starting Out—Why Education for Refugees Matters. www.unhcr.org/starting-out-why-education-refugees-matters.
UNU. (2023). Refugee Settlements Extremely Vulnerable to Climate Crises. https://unu.edu/article/refugee-settlements-extremely-vulnerable-climate-crises#:~:text=Refugees%20residing%20in%20Kenya%2C%20Rwanda,high%20temperatures%20are%20observed%20increasing.
Wahlbeck, H., Kvist, L. J., and Landgren, K. (2020). Art Therapy and Counseling for Fear of Childbirth: A Randomized Controlled Trial. *Art Therapy* 37, 123–130. doi:10.1080/07421656.2020.1721399.
WFP. (2023). Horn of Africa Hunger Crisis Pushes Millions to the Brink. www.wfp.org/stories/horn-africa-hunger-crisis-pushes-millions-brink#:~:text=Millions%20of%20people%20in%20the,Programme%20(WFP)%20warned%20today.
WFP USA. (2024). Hunger in the 4 Largest Refugee Camps in the World. www.wfpusa.org/articles/hunger-largest-refugee-camps-world/.
WHO. (2022). Arts and Health: Supporting the Mental Well-being of Forcibly Displaced People. https://cdn.who.int/media/docs/librariesprovider2/country-sites/who_arts-and-health—forcibly-displaced-people-(final).pdf?sfvrsn=2800af42_1&download=true

9

PERFORMATIVE PEDAGOGY AND AFFECTIVE JUSTICE

The Embodied Self and the Body Politic

Fiona Woods

Introduction

This chapter begins with a consideration of that systemic violence, moving on to an examination of aesthetics as a political and pedagogical force. It highlights the ways in which aesthetics is bound to both oppressive and liberatory formations, particularly in the context of coloniality (Quijano, 2008).[1] The value of decolonial thinking for this analysis lies partly in its refusal to be bound by aesthetical, ontological and epistemological values that have their origins in the colonial/capitalist imaginary of the West. Decolonial theory therefore offers multiple tools for challenging the 'capitalocentric modes of thought that have crowded out narratives of care and social justice in thinking about social change' (Lynch, 2022, p. 3). These are some of the ideas that influence my approach to pedagogy, particularly in the design of Territories of the Self, which encourages a collective engagement with matters related to social power, spatial justice and the norms inscribed on bodies. This is elaborated in greater detail in the second half of the chapter.

Systemic Violence

Capitalism is an inherently violent system, founded on pre-existing patriarchal and racial hierarchies of power and wealth, which it legitimates through forms of institutionalisation and cultural hegemony. As a system, capitalism co-opts the State to act as an agent for capital, to regulate in favour of the 'free' market, and to protect its interests by legal, military and imperialist means (Harvey, 2005). The current order of extractive capitalism is referred to here as the Capitalocene (Moore, 2017). While Anthropocene is more commonly used, it arguably proposes a false universalism in relation to

DOI: 10.4324/9781003456636-10

the origins and impacts of socio-ecological devastation. The concept of the Anthropocene is envisioned by means of what Donna Haraway describes as the 'conquering gaze from nowhere' (Haraway, 1988, p. 581). Kathryn Yusoff contends that the term obscures the 'structures of thought and material arrangements that brought the Anthropocene into being in the first place' (Yusoff, 2019). She historicises 'white extraction regimes' as originators of the current social and ecological crisis. Many theorists argue that the violence of colonial extraction, including the extraction of labour from enslaved bodies, is foundational to capitalist accumulation (Federici, 2017; Quijano, 2008; Yusoff, 2019; Rice, Long, and Levenda, 2022). Capitalocene has its limitations as a term, but it points to the systemic roots of devastating climate change, the 'nexus of capitalism-colonialism-patriarchy that set fire to the planet in the first place' (Demos, 2019).

The aesthetic and affective conditions of the Capitalocene are central to this analysis. Its systemic violence can range from brutality, murder and expulsion at one end of a spectrum (Sassen, 2014), to the alienation of affective life at the other. Media technologies act as a channel for what Hito Steyerl describes as a 'dictatorship of affect and noise' (Steyerl, 2007). A ubiquitous techno-aesthetics fuses sex, violence and a state of emergency to create a constant atmosphere of heightened alert. Glamour, titillation and fear are conflated in a kind of 'pornography' that acts on the nervous system of populations (Steyerl, 2007). It is a characteristic of contemporary power as such, says Steyerl, that it can trigger, channel, mediate and market sensation: 'power operates … within the senses' (Steyerl, 2007).

In these conditions, politics is not only aestheticised but is 'exercised as aesthetics' (Steyerl, 2007). Intrusive and addictive media technologies operate as tools for a social ordering of perception, fashioning ways of looking, tasting, feeling, moving and smelling that can be instilled into the postures and rhythms of our bodies. Those technologies manufacture and exploit an attention economy, 'a political economic rationale for reconstituting capitalism in the locus of the body' (Crogan and Kinsley, 2012, p. 3). What is at stake here is the power to calibrate the discriminative capacities of populations by shaping what we sense and how we make sense of it, which is an aesthetic process. If, as James Williams argues, attention is 'the full stack of navigational capacities across all levels of human life' (Williams, 2018, p. 149), then conditions that support people to focus and refocus their attention are crucial for the development of individual and collective agency. This is what situates aesthetics at the core of politics (and vice versa); aesthetics is a terrain of action and a set of processes by which to identify, analyse, reframe and (de)construct the regimes of sensibility and intelligibility that establish the self-evident 'facts' of a social order.

Disciplining Sensibilities

Modern Aesthetics is a versatile concept, capable of speaking to a wide range of preoccupations, including 'freedom and legality, spontaneity and necessity, self-determination, autonomy, particularity and universality' (Eagleton, 1990,

p. 3). It emerged as a theory of sensibility, sensation and sentiment in eighteenth-century Germany to counter a 'rationalist devaluation of sense cognition' that permeated Enlightenment philosophy (Gregor, 1983, p. 369). Not especially concerned with art *per se*, aesthetics was first 'born as a discourse of the body' (Eagleton, 1990, p. 13). Due to a 'certain indeterminacy of definition' (Eagleton, 1990, p. 13), its internal contradictions continue to lend themselves to diverse theoretical and political positions.

One of the key texts of early modern aesthetics, Friedrich von Schiller's *Letters on the Aesthetical Education of Mankind* (1794), proposed the cultivation of 'an aesthetic disposition of the mind', as a basis for the transformation and liberation of 'mankind'. This 'aesthetic modulation of the psyche' (Schiller, 1794, letter XXIII) could be achieved by bringing the 'sensuous instinct' and the 'formal instinct' together in the 'instinct of play' (Schiller, 1794, letter XIV). Beauty would be an educational experience, arising from the 'free play of our understanding and imagination'. This education would give rise to an 'aesthetical state of mind' (Schiller, 1794, letter XXIII), a harmonious interaction between understanding and imagination that was essential for a truly free state of mind. For Jacques Rancière, this work by Schiller amounted to a 'manifesto' for a new sensory regime (Rancière, 2004, p. 24); Terry Eagleton also recognised in Schiller's *Letters* a programme for disciplining the sensibilities of the newly emerging political subject of the bourgeois revolution (Eagleton, 1990).

Similarly, an aesthetic modulation of the psyche was enacted on the lower classes in the eighteenth and nineteenth centuries, according to Walter Benjamin, albeit of a different kind. Wrought by the physical and psychic shocks of industrialisation and militarisation (Buck-Morss, 1992, p. 16), the 'synaesthetic system' of the human was forced to take on a desensitising role, to block out the sensory violence of daily life. Susan Buck-Morss describes this sensorial alienation as an 'anaesthesia', a shadow-twin to the 'privatized fantasy world that functioned as a protective shield for the senses and sensibilities of [the] new ruling class' (Buck-Morss, 1992, p. 22). Karl Marx argued that this anaesthesia followed inevitably from the logic of private property, 'the dismal displacement of our sensuous plenitude onto a single drive to possess' (Eagleton, 1990, p. 199). In this sense, Marx is 'profoundly aesthetic' (Eagleton, 1990, p. 199) because he seeks to restore to the body the sensuousness and powers that have been estranged and misappropriated by Capital:

> not only the five senses but also the so-called mental senses, the practical senses (will, love, etc.), in a word, *human* sense, the human nature of the senses, comes to be by virtue of *its* object, by virtue of *humanised* nature.
>
> *(Marx, 1844)*

Colonial (An)aesthetics

A profound sensorial alienation was also effected on populations across the global south, from the sixteenth century onwards, as the sensory regimes of European colonisers were imposed on indigenous peoples through violence, religion, education and other institutional arrangements. 'The physical violence of the battlefield', wrote Ngũgĩ wa Thiong'o, 'was followed by the psychological violence of the classroom' (Ngũgĩ, 1986, p. 9). In the early decades of the twentieth century, movements for independence across the colonised world developed radical, anticolonial pedagogies, including the educational work of Rabindranath Tagore in India and Pádraig Pearse in Ireland, at a time when both countries were fully under British colonial rule. Tagore and Pearse were literary figures who were familiar with one another's work.[2] They were both active in cultural revival movements as part of a drive for independent statehood. They shared a commitment to the restoration of their native language (Bengali and Gaeilge respectively) as the primary medium for education, in order to depose the cultural hegemony and values of Anglicisation.

As Tagore, Pearse and many others recognised, the sensuous nature of language makes it a fundamental carrier of indigenous sensibilities. Each language has its own musicality, its own situatedness in the body in terms of its modes of articulation and the sounds that it makes. A language is founded on ontological and cosmological suppositions that inflect its ways of framing the world. In *Decolonising the Mind: The Politics of Language in African Literature* (1986), Ngũgĩ emphasises the significance of language as a site of decolonisation: 'language as culture is the collective memory bank of a people's experience in history' (Ngũgĩ, 1986, p. 15). The imposition of a colonial language leads to a 'disassociation of the sensibility of [a] child from his [or her] natural and social environment, what we might call colonial alienation' (Ngũgĩ, 1986, p. 17).

When Pearse said 'Tír gan Teanga, Tír gan Anam' ('a country without a language is a country without a soul'),[3] he was echoing a central tenet of cultural nationalist theorists across the nineteenth century, such as the German, Johann Herder, who wrote: 'A people … has nothing dearer than the language of its fathers. Its whole spiritual wealth of tradition, history, religion and all the fullness of its life, and all its heart and soul lives in it' (Herder, n.d., in Coolahan, 2016, p. 19).

While Pearse was an active nationalist, Tagore viewed nationalism as a European construct, 'at odds with the syncretic Indian civilization, which had always been pluralistic and open to intercultural exchange and understanding' (Banga, 2023, p. 617). Tagore's ambition was to go beyond the nation-state by 'conceiving of "Asia" as a new form of 'imagined community' (Shani, 2022, p. 3). Georgio Shani shows how 'Asia', for Tagore, constituted a 'method' of decolonising local imaginaries, 'whereby different cosmological traditions can learn to converse with another without referring to the West' (Shani, 2022, p. 3).

Having suffered in the colonial education system themselves, Tagore and Pearse shared an interest in decentralised, progressive models of education. Pearse had a child-centred philosophy; 'the main object in education' he stated, 'is to help the child to be his own true self' (Pearse, 1952, p. 28). Tagore's approach had three basic tenets: 'freedom for the learner, creation of an environment that enables the student to develop a healthy kinship with nature, and cultivation of the pupil's creativity or imagination' (Quayum, 2016, p. 4). The primary aim of his approach to education was encapsulated in his use of the term *AtmaShakti*, which can be translated as 'strengthening of the soul' (Banga, 2023, p. 617). Pearse likewise anticipated that a liberating education would 'foster the elements of character native to the soul, to help bring these to their full perfection' (Pearse, 1952, p. x). There are echoes here of Schiller's views on 'the beautiful soul' in which 'sensuousness and reason, duty and inclination are in harmony, and grace is their expression as appearance' (Schiller, 1793, p. 368).[4]

Whether or not Tagore or Pearse were familiar with Schiller's work, the humanist and Enlightenment principles that inform *Letters on the Aesthetical Education of Mankind* (Schiller, 1794) are identifiable in their pedagogies, particularly in their conception of education as an aesthetical process. Their approaches, while learner-centred, were nonetheless oriented towards the cultivation of sensibilities appropriate for a post-colonial condition. Like Schiller, these new sensibilities were to be achieved through the effects of beauty and play; this manifested as an emphasis on indigenous language, on theatre and literature, and on embodied forms of learning such as gardening, dance and sport. The embodied sensorium of the student was fully engaged to liberate the senses from the (an)aesthetics of anglicisation.

Decolonising Aesthetics

Rooting out the bodily foundations of coloniality has been asserted in scholarship, literature and cultural work across the colonised world, from the late nineteenth century to the present day. The embodied sensibilities of colonised people have been violated over many centuries, leaving an open 'colonial wound' that must be healed (Mignolo and Vazquez, 2013). In their essay *Decolonial AestheSis: Colonial Wounds/Decolonial Healings* (2013), Walter Mignolo and Rolando Vazquez argue that Modern Aesthetics, in the form discussed earlier, was integral to the structures of imperial control that began to emerge with the establishment of 'the Atlantic commercial circuit and the colonization of the New World', from the sixteenth century onwards (Mignolo and Vazquez, 2013). They use the critical formulation 'modern aestheTics' to refer to this 'phenomenon of a modern/colonial … normativity that colonized the senses' (Mignolo and Vazquez, 2013). The false universalism of 'modern aestheTics' sought to overwrite geopolitical variations 'of sensing, knowing and believing' (Mignolo and Vazquez, 2013). 'Decolonial aestheSis', on the other hand, names critical positions and artistic practices 'that aim to decolonize the senses, that is, to liberate them from the regulations of modern, postmodern, and altermodern aestheTics' (Mignolo and Vazquez, 2013).

The internal contradictions of aesthetics can be seen at work here. Its origins as a programme of sensorial conditioning in the nexus of capitalist-colonialist power make it a key mechanism by which to situate ideology in human bodies; 'structures of power must [first] become structures of feeling' (Eagleton, 1990, p. 21). However, as Eagleton has also argued, the properties of self-determination and autonomy that are demanded by the aesthetic assert a space from which predetermined interests are absent. For Christoph Menke, the aesthetic is:

> a dimension, a dynamic, a force of the soul, and with that, a source of everything that makes us who we are—in contrast to theory, because it has neither subject nor content; in contrast to praxis, because it has no goal; in contrast to concept, because it has no rules; in contrast to society, because it has no norms; in contrast to individuality, because it has no owner.
>
> *(Menke, 2015, p. 41)*

Menke's neo-Kantian formulation sees in aesthetics a non-teleological suspension of established functions and relations, which opens a fissure in the connection between the sensuous and the sensible. That fissure, following Rancière, is the site for a decentring and disentangling of the normative and habitual relations of sense-making (Rancière, 2010).

The evolving field of decolonial aesthetics, informed by the work of theorists and artists in Latin America, Eastern Europe, the Caribbean, Asia, Australia and Aotearoa New Zealand is generating vital tools for examining and resisting the ongoing violence embedded in the psychic, physical and social infrastructures of coloniality and the Capitalocene.

> The goal of decolonial thinking and doing is to continue re-inscribing, embodying and dignifying those ways of living, thinking and sensing that were violently devalued or demonized by colonial, imperial and interventionist agendas as well as by postmodern and altermodern internal critiques.
>
> *(Transnational Decolonial Institute, 2011)*

Without seeking to collapse the distinction between the colonisation of civilisations by European powers and the ongoing coloniality of the Capitalocene, there are correspondences in the alienating effects of patriarchal, capitalist neocolonialism on the sensibilities of people everywhere. A 'woundedness' is evident in the mental health crisis that affects young people in particular. This is often attributed to 'social media', as a kind of abstract force. However, social media is an expression of the Capitalocene's relentless drive towards a commodification of everything, including the sensuous, the personal and the relational. Social media delivers an alienated sociability directly into the psychic space of its users. Supporting young people to develop ways of resisting the violence of that commodification is one of the challenges that my approach to pedagogy tries to address.

Performative Pedagogy

Territories of the Self (TOTS) is a 10-week seminar course that is part of undergraduate curriculum in the School of Art & Design at the Technological University of the Shannon in Limerick. Up to 30 art and design students select the course from several courses on offer. The course was initially designed for second year students who had spent their first year of college in COVID-19 lockdown. Through a combination of conceptual and experiential learning, TOTS supports students to test, and to trust, their capacity to respond critically and purposefully to conditions that they encounter in the world. Individual and collective processes of meaning-making emerge through reading and written reflection, aesthetic works, and performative actions in public or semi-public space. A formal session takes place once a week, lasting an hour and a half, with semi-structured independent learning taking place between sessions. The formal session operates as a focal point, a collective engagement with concepts and aesthetic matters related to social power, spatial justice and the norms inscribed on bodies.

A performative approach to pedagogy recognises that 'learning can take place through our entire bodies; knowledge acquisition can be multisensory' (Campbell, 2020, p. 7). Critical reflexivity is built through a 'full body engagement' as described by Elyse Lamm Pineau; 'information must be engaged somatically as well as intellectually. It is the dialectical process of doing and reflecting, experiencing and interpreting' (Pineau, 1994, p. 17). The manner of learning in performative pedagogy is itself epistemological; the means by which students come to know is purposefully somatic and intellectual. Epistemological 'authority' is distributed through a rigorous-but-intimate process of embodied, cognitive inquiry. Students become aware of the 'negotiation and enactment of possible knowledge claims' (Pineau, 1994, p. 10). As Lee Campbell argues, 'performative pedagogy creates a bridge between the mind and the body in terms of the embodied experience and cognitive process; how meaning making takes place' (Campbell, 2020, p. 14).

The title of the course situates the 'self' at the heart of the exploration, indicating to students that embodied and emotional forms of knowledge will form part of their research, alongside more abstract concepts. Prompts are employed to encourage students to pay close attention to their sensory experience and, through specific actions, to observe shifts between different states of perception, awareness and analysis. The very first action of the course asks students to sit with closed eyes and, using a blind contour drawing process, to map 'the edges of themselves' in real-time and space.

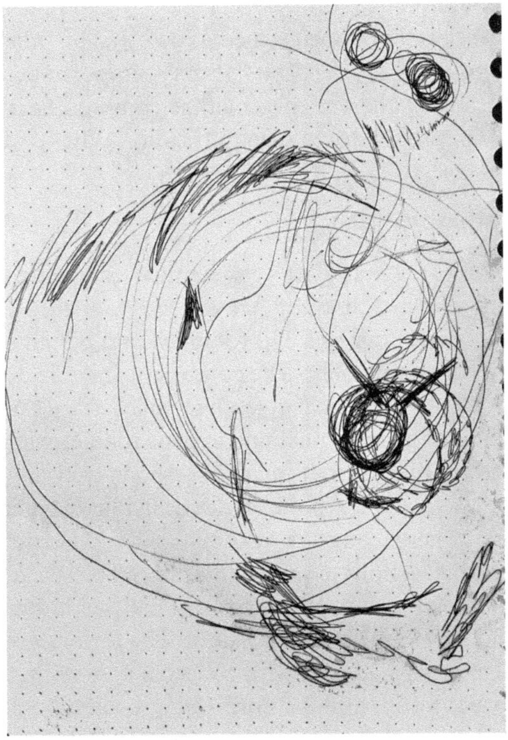

FIGURE 9.1 'Map' your bodily field. L.K., blind contour drawing, 2024. 'While drawing this I tried to focus in on the sounds around me, both in and out of the class and tried to distinguish how far away they were from me. I also wanted to illustrate on how my body felt on the inside, such as capturing the nerves and the senses, mainly the sense of touch and sound.'

In this way, their bodily and affective realities are positioned as the ground for subsequent practices of meaning-making.

By using the body as material, process and site, students learn to rewrite and re-present the cultural codes inscribed on their bodies and, in doing so, to construct their own identities and create new cultural myths with which to challenge the body politic (Garoian, 1999, quoted in Campbell, 2020, p. 4).

This initial emphasis on the embodied self is gradually expanded to encompass a critical engagement with the body politic. Insights gained through playful engagements with a range of environments are progressively mapped onto observations about the ways in which environments 'perform' the social order. Students share their research on a digital platform, which they update each week to encourage peer learning and promote a collective epistemology.

Spatial Sensibilities

The course begins with two readings, *Species of Space* (Perec, 1999) and 'I hear sounds inside my head' (Rajkowska, 2020). Both reflect, in different ways, on the transformation of spatial sensibilities. Georges Perec's work draws attention to the 'infra-ordinary', the things that go without saying in our occupation of domestic space. He emphasises the spaces between things, the non-spaces of lived experience that are generally perceived as 'empty'. He does so in a playful and poetic way that gently introduces Henri Lefebvre's more complex ideas about the social production of space (Lefebvre, 1991). Joanna Rajkowska writes from the perspective of an artist who makes work in public space. Her ideas about public space are informed by two traumas. As a child, her mother suffered from progressive psychosis. Whenever they were out in public Rajkowska would have to rapidly interpret the environment to avoid places that might trigger her mother's intense anxiety and paranoia. She learned to 'react like a seismograph … to sense the energy of places' (Rajkowska, 2020, p. 200). Those places, she realised later, were also saturated with 'post-communist … historical trauma' (Rajkowska, 2020, p. 204). Spaces that might be perceived as empty or psychologically neutral for some people were laden with threat, individually for her mother, and collectively for the citizens of that 'post-communist realm' (Rajkowska, 2020, p. 204).

The two readings are coupled with prompts for embodied actions, in the tradition of instructive art:

- 'Map' the edges of yourself (your bodily field).
- 'Map' every bed you have ever slept in.
- Occupy a mundane space in a way that is 'poetic' (following Perec).
- Working in small groups, find a 'useless' space and occupy it in an unexpected way.
- Listen to [selected audio-work] and follow its storyline in real time to navigate through a familiar or unfamiliar location.

The actions take students through an analysis of personal space, psychological space, instrumentalised space, and socially constructed spaces. Together as a group we look at video essays and artist films that respond to Perec's text. We discuss 'territory' as a nested phenomenon. We study 'play' as an upending of spatial norms. Together, the readings, discussions and actions alert students to different spatial sensibilities. Gestures and postures of habitation and occupation are amplified. Space becomes a bodily and aesthetic experience, a process of coming-into-being, susceptible to distortion, subversion and mutation.

Omar Rivera argues that 'decolonizing space is, effectively, a corporeal transformation, a modulation of physically finding one's bearings within

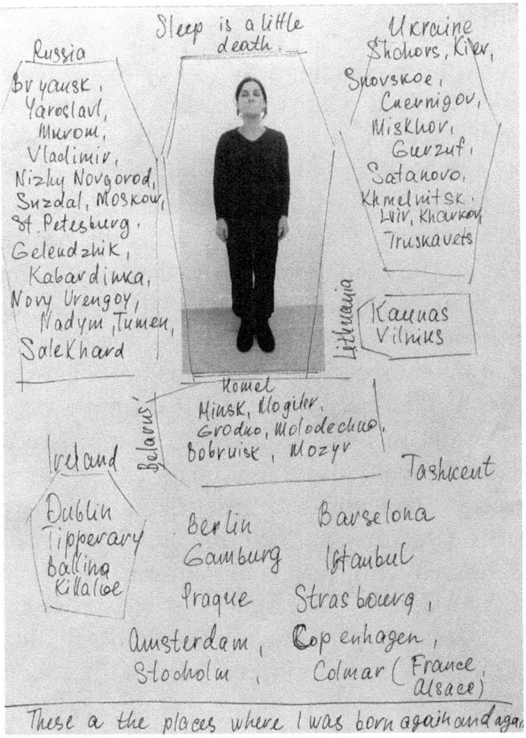

FIGURE 9.2 'Map' all the beds you have ever slept in. A.S., *Sleep is a little death …* Collage/drawing, 2024.

spaces layered by colonial histories and socialities that determine one's perceptual and affective habits' (Rivera, 2022, p. 145). His analysis of 'complex, situated junctures of anticolonial resistance', in a range of contexts both historical and geopolitical (Rivera, 2022, p. 3), refuses a binary of resistance versus passivity. Arguing that Andean indigenous resistance (the focus of his analysis) cannot necessarily be recognised as such within the frame of colonial oppression, Rivera proposes that resistance (analytically distinct from 'liberation') attends to the ways in which bodies have been disciplined by power. It is 'physically metamorphic … connected to a sentience that is in-stilled while in transformation' (Rivera, 2022, p. 3). Bodies in resistance are bodies in flux, oriented and disoriented, impure, unruly, destabilising.

The first part of the course is therefore about developing a critical relationship with one's own situatedness in social and physical architectures. Working through affective and bodily encounters, students observe the emergence of unfamiliar 'postures, affects and perceptions' (Rivera, 2022, p. 3) as they engage with familiar socio-spatial formations, working at the edge of their comfort zone.

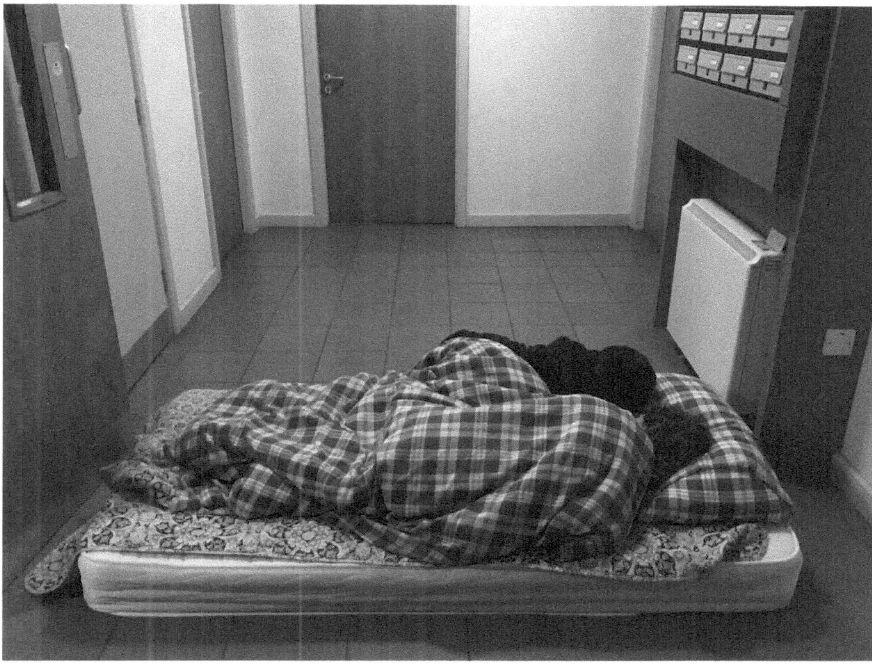

FIGURE 9.3 Find a 'useless' space and occupy it in an unexpected way. A.S., *action*, 2023.

The Body Politic

A wider politics of spatial justice informs the second part of the course, which emphasises the marginalisation of vulnerable bodies in public space. The pivot from embodied self to body politic is anchored in questions about transformative and disability justice. Three readings are selected as coordinates for the intellectual work: 'Participatory Mapping of Disabled Access' (Kitchin, 2002); 'The Hidden Ways Urban Design Segregates the Poor' (Jaffe, 2014); and *Building paranoia: the proliferation of interdictory space and the erosion of spatial justice* (Flusty, 1994).

In Ireland disability legislation is 'tokenistic, weak and very poorly enforced' (Kitchin, 2002: 46).[5] The agency of disabled people is often overlooked or undermined. Rob Kitchin's reading offers a critical analysis of these conditions and shows how they informed a collaboration between academic researchers and a pan-disability advocacy organisation, the Newbridge Access Group [NAG] in 2002. The Newbridge Participatory Mapping Project trained members of NAG to map restrictions and obstacles to access and mobility in their town. As well as carrying out the mapping themselves, participants made decisions about the priorities of the project, as well as the representational language of the resulting maps. This approach, informed by 'feminist praxis', challenged the usual power-relations embedded in academic research 'about' disabled people (Kitchin, 2002, p. 45).

The reading offers a rigorous but accessible introduction to the social model of disability and the ableist gaze. Many students have a disability themselves or know someone who does. I have found that students are relatively quick to acknowledge their own 'ableist gaze', and to grasp how that creates an exclusive spatial order. By analysing what is invisible to the ableist gaze, our discussions extend the idea of 'the gaze' to other physical and social architectures of oppression. We explore how the needs and experiences of girls in public space are disregarded, for example, and how disciplinary and hostile architectures, both physical and social, are used to expel the homeless and the poor from the public realm. The purpose of these discussions is to study how institutional structures and social arrangements devalue certain bodies, and how that devaluation translates into socio-spatial structures. We also study works that contest such spatial oppressions.

Public space in Ireland is heavily policed by state and corporate interests. Risk-aversion has been worked into the grain of the social imaginary. The second set of prompted actions are situated in public or semi-public space and encourage students to step further out of their comfort zones.

- Working with a safety partner, walk through a public or semi-public 'territory' with your eyes closed, for as long as you can. Notice how your bodily field is impacted by different sounds and sensations. You will probably feel vulnerable; pay attention to that experience.

Without exception, students report that walking with closed eyes in public space is transformative. As well as a vastly increased sensitivity towards sound, smell and temperature, they describe issues with balance and proprioception, feeling as though gravity is stronger, their limbs are bigger, the ground is rougher. They report a heightened awareness of their own and other presences in the 'territory'. The experience of vulnerability is shocking for many people; some of them describe relaxing into it, while others find it threatening.

The next action asks students to consider the mechanisms of control that saturate public and semi-public space, and the social pressure to conform to those controls.

- Working individually or in small groups, take a 15-minute walk in the city. Map the 'signs' that you encounter. Pay attention to how they operate to regulate your behaviour. Enact a small refusal. Create a short video essay reflecting on those conditions.

Most students report that the sheer volume of directive signage surprises them when they pay attention to it. They also find that acts of (even minor) disobedience are very difficult to do. During the formal sessions they present

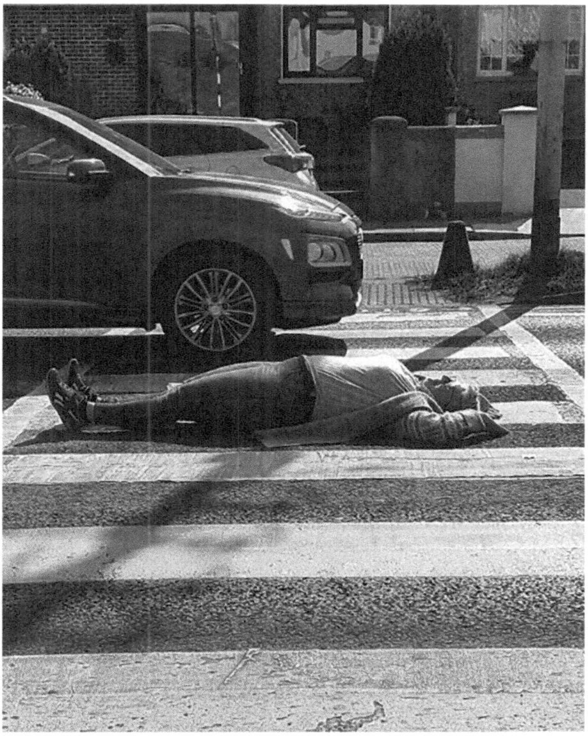

FIGURE 9.4 Transgress or subvert a boundary or condition, using poetic strategies. T.O'C., public action, 2022.

their video essays and engage in discussions, drawing on different modes of analysis prompted by the readings. The breadth of response to the brief is notable. Because students bring different states of perception, awareness and analysis into play when they carry out their actions, the tone of their work may be political, emotional, satirical, intellectual, phenomenological, artistic, or a combination of these.

Sentience and Resistance

The final prompted action for the course is as follows:

- Working individually or in small groups, identify an aspect of public space that inhibits full public access. Transgress or subvert that boundary or condition, using poetic strategies.[6]

Transgression and subversion carry a higher social risk than an act of 'refusal', as proposed in the previous action. The purpose of this intensification is to encourage students to find the edge of what they are willing to conform to, with a view to understanding the stakes of their own agency. Before the actions take place, we discuss how uncomfortable it can be to transgress social norms. Students are assured that they do not need to go past a point of discomfort that they can tolerate. They are strongly advised not to put themselves in harm's way. The resulting actions have therefore ranged from complimenting strangers, to offering them sweets on a bus, to occupying restricted spaces, to dancing in a public space, and many things in between.

When students present their actions to the group, we use Harold Garfinkle's work on 'breaching experiments' (Garfinkle, 1967) as a basis for unpacking their own views on the power of social norms.

The emphasis on poetic strategies in the prompt goes back to Rivera's ideas about 'resistant embodiments' (Rivera, 2022, p. 38). Arising from 'a sensibility informed by Latin American and Latinx aesthetics' (Rivera, 2022, p. 1), as well as Andean epistemologies and cosmologies, Rivera elucidates a form of anticolonial resistance that arises and persists through communal, aesthetic practices. This form of resistance may not be 'intelligible' in the register of colonial oppression/liberation, but, according to Rivera, its modes of 'sense perception, affects, postures, and other bodily enactments … decenter modern/colonial elucidations and inhabitations of social formations, even liberatory ones' (Rivera, 2022, p. 3). This is resistance that begins at the level of 'physical sentience' (Rivera, 2022, p. 15), that is to say, a physical

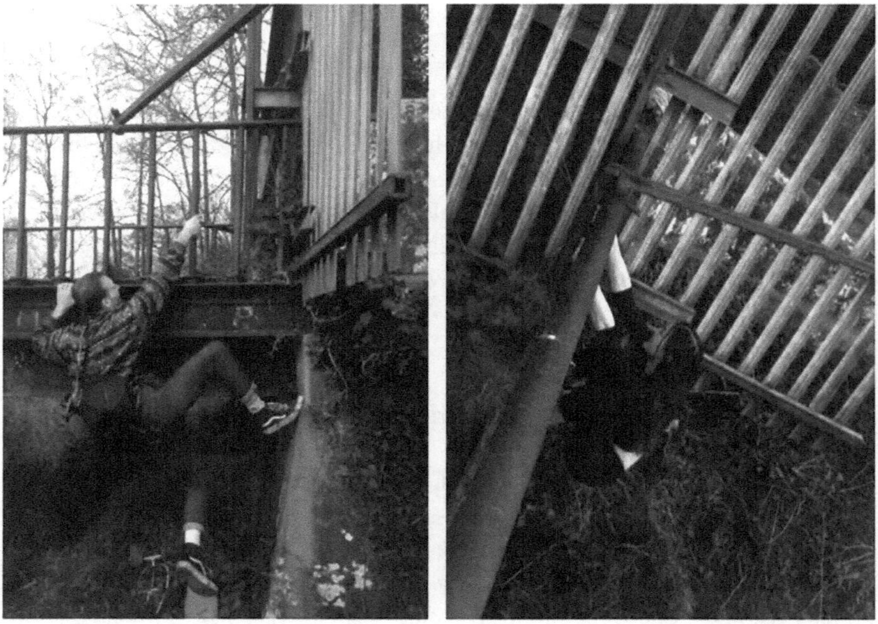

FIGURE 9.5 Transgress or subvert a boundary or condition, using poetic strategies. S.K., details of public action, 2022.

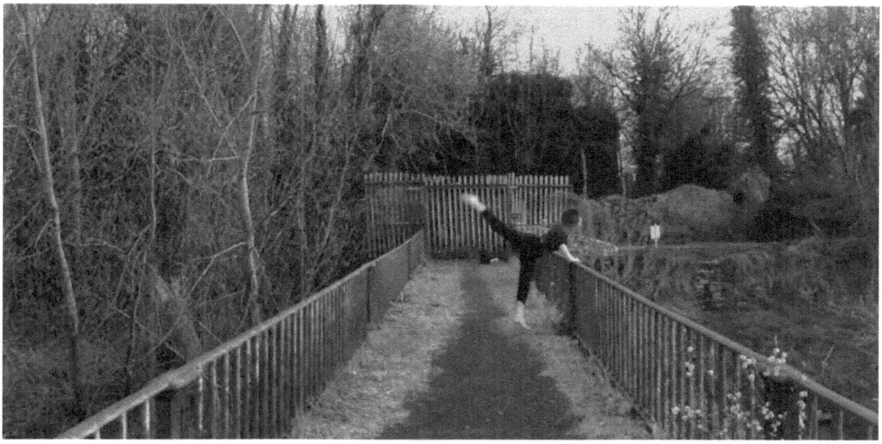

FIGURE 9.6 S.K., detail of still from video work, riverbed borders, 2022.

attentiveness to the social conditions and bodily dispositions that shape not only what we know, but how we come to know. Decentring normative and habitual ways of inhabiting, responding to and communicating about the capitalocentric world is one of the strategic objectives of the course, as a basis for challenging the normalisation of systemic violence. One student's experience reflects this:

> After [the seminar course] I felt as if a colourblind person was given the opportunity to look at the world through the eyes of ordinary people, my emotional shock was so strong. Ordinary affairs and even drawing lost their significance at that time, since it took me time and internal resources to rethink and accept seemingly obvious things.
>
> *(A.S., 2024, email to the author)*

Conclusions

The neoliberal capture of educational imaginaries and institutions proceeds apace. Kathleen Lynch argues that 'the primary purpose of education and welfare institutions [in the context of neoliberal capitalism] is seen as conditioning and training individuals to be self-reliant, individually responsible and entrepreneurial' (Lynch, 2022, p. 2). An increasingly technocratic understanding of teaching as 'instructional practice' jeopardises its intellectual, creative, and radical possibilities. I share Lynch's view that the profound harms caused by neoliberal capitalism can only be challenged by moving 'beyond the non-relational, self-referential ontology that underpins neoliberalism's culture and politics' (Lynch, 2022, p. 2). To reposition 'caring and affective justice' at the heart of the social order (Lynch, 2022, p. 2) it is necessary to de-normalise the standpoint from which the Capitalocene makes sense. One of the ways that I can contribute to that, as a teacher, is by creating opportunities for students to examine 'socially conditioned patterns of affective tendencies and disposition' (Suša et al., 2021), to consider how those patterns might be lodged in their own embodied experiences, and to make choices about the forms of social conditioning that they are willing to tolerate and accept.

There is a view that artists and designers are the perfect neoliberal subjects, 'entrepreneurs of themselves' (ARTmargins, 2021), converting their subjectivity into creative capital, competing for visibility, etc. I propose another view, namely that art and design education can help students to 'engage with discomfort and uncertainty without feeling overwhelmed or irritated ... and to sit with internal complexities, paradoxes and contradictions' (ARTmargins, 2021). These are vital capacities for projects of world-making, in which some shattering of existing worlds may be necessary. One of the values of third level art and design education is that it relies on freedom for both students and teachers to experiment, to contest and to dissent. Expertise is both acknowledged and constantly upended, with epistemological parameters subject to radical reworkings by students. As a teacher I view that as a privilege and a responsibility.

I concur with Jack Halberstam's account of study as 'a mode of thinking with others separate from the thinking that the institution requires of you' (Halberstam, 2013, p. 11). I try to think with students and to let myself to be changed by them. At the outset of every course, I draw attention to the power imbalance that exists in the classroom. Although acknowledging that imbalance does not change it, it creates a starting point for exploring a redistribution of epistemic authority amongst us. Finding ways to respond critically and purposefully to the systemic violence of the Capitalocene is something that we are learning to do together.

Notes

1 The idea of coloniality emerged through a series of events and discussions in the Americas between 1998 and 2008 (Mignolo, 2011). A key architect of the theory, Anibal Quijano, identifies it as a relational concept, an index of the invisible dark side of modernity. 'Coloniality is shorthand for the colonial matrix of power ... [which] is, in short, the pillars on which Western civilization rests' (Mignolo, 2011).

2 In 1913, W. B. Yeats arranged the staging of Tagore's play *The Post Office*, alongside Pearse's play *An Rí*, at the Abbey Theatre in Dublin. This was the first presentation of Tagore's theatrical work outside of India. Tagore also staged *An Rí* with pupils in his school at Santiniketan in Bengal in 1915.

3 This was the motto of the Gaelic League, a nineteenth-century organisation established to 'revive' Gaelic language and culture which had fallen out of use across the island.

4 Western theories of education are generally grounded in rationalist philosophies in which the 'soul' is no longer a reference point. A non-religious interpretation of 'soul' in the aforementioned pedagogical positions might be a purposeful, individual wholeness, in harmony with its environment, a rich form of agency extending beyond the individual.

5 It is worth noting that very little has changed in terms of disability justice in Ireland in the 20 years since Kitchin wrote the article.
6 I offer an avowedly rough definition of poetics as a deliberate spacing of 'forms' to generate rhythms/affects, as well as how those are experienced and interpreted, and/or as ways of 'bringing meaning into being through making' (Drucker, 2013, p. 82).

References

ARTmargins. (2021). Art Under Neoliberalism, Roundtable. *ARTMargins* 10(3), 126–158. http://direct.mit.edu/artm/article-pdf/10/3/126/2034986/artm_a_00303.pdf (accessed 7 April 2024).

Banga, S. (2023). The Global Relevance of Tagore's Cosmopolitan Educational Philosophy for Social Justice in a post-Westphalian World. *Journal of Philosophy of Education* 57(3), 611–625. https://doi.org/10.1093/jopedu/qhad051 (accessed 13 December 2023).

Benjamin, W. (1935). The Work of Art in the Age of Mechanical Reproduction. In H. Arendt (ed.), *Illumination*. Translated from German by H. Zohn. New York: Schocken Books.

Buck-Morss, S. (1992). Aesthetics and Anaesthetics: Walter Benjamin's Artwork Essay Reconsidered. *October* 62, 3–41.

Campbell, L. (ed.). (2020). *Leap into Action: Critical Performative Pedgaogies in Art & Design Education*. New York: Peter Lang.

Coolahan, J. (2016). Pádraig Pearse the Educationalist. In B. Mooney (ed.), *Education Matters Yearbook, 2016–2017*, pp. 17–24. https://educationmatters.ie/wp-content/uploads/2017/02/Padraig-Pearse-the-Educationalist.pdf (accessed 13 December 2023).

Crogan, P. and Kinsley, S. (2012). Paying Attention: Towards a Critique of the Attention Economy. *Culture Machine* 13, 1–29. https://culturemachine.net/wp-content/uploads/2019/01/463-1025-1-PB.pdf (accessed 9 August 2016).

Demos, T. J. (2019). The Agency of Fire: Burning Aesthetics. www.e-flux.com/journal/98/256882/the-agency-of-fire-burning-aesthetics (accessed 13 December 2023).

Drucker, J. (2013). Diagrammatic Writing. *New Formations* 78, 83–101.

Eagleton, T. (1990). *The Ideology of the Aesthetic*. Oxford: Blackwell.

Federici, S. (2017). *Caliban and the Witch: Women, the Body and Primitive Accumulation*. Autonomedia.

Flusty, S. (1994). *Building Paranoia: The Proliferation of Interdictory Space and the Erosion of Spatial Justice*. Los Angeles Forum for Architecture and Urban Design.

Garfinkle, H. (1967). *Studies in Ethnomethodology*. Prentice-Hall.

Gaztambide-Fernández, R. (2014). Decolonial Options and Artistic/AestheSic Entanglements: An Interview with Walter Mignolo. *Decolonization: Indigeneity, Education & Society* 3(1), 196–212.

Gohain, H. (2011). Two Roads to Decolonisation: Tagore and Gandhi. *Economic & Political Weekly* xlvi(31), 23–26.

Gregor, M. J. (1983). Baumgarten's 'Aesthetica'. *The Review of Metaphysics* 37(2), 357–385.

Halberstam, J. (2013). The Wild Beyond: With and For the Undercommons. In S. Harney and F. Moten (eds), *The Undercommons: Fugitive Planning & Black Study*, pp. 2–12. www.minorcompositions.info/wp-content/uploads/2013/04/undercommons-web.pdf (accessed 11 April 2024).

Haraway, D. (1988). Situated Knowledges: The Science Question in Feminism and The Privilege of Partial Perspective. *Feminist Studies* 14, 575–599.

Harvey, D. (2005). *A Brief History of Neoliberalism*. Oxford: Oxford Academic.

hooks, b. (1994). *Teaching to Transgress; Education as the Practice of Freedom*. New York: Routledge.

Hundorova, T. (2023). How Peripheries Talk Amongst Themselves — Or Ukraine, Eurocentrism and Decolonization, in Krytyka, Thinking Ukraine. https://krytyka.com/en/articles/how-peripheries-talk-amongst-themselves-or-ukraine-eurocentrism-and-decolonization#footnote-14 (accessed 12 December 2023).

Jaffe, E. (2014). The Hidden Ways Urban Design Segregates the Poor. www.fastcompany.com/3034206/the-hidden-ways-urban-design-segregates-the-poor (accessed 14 March 2025).

King, A. (2018). Tagore and Yeats: India-Ireland Relationship Goes beyond the Political. www.tcd.ie/trinitylongroomhub/media/news/articles/2018-10-26-Sirshendu-Majumdar.php (accessed 13 December 2023).

Kitchin, R. (2002). Participatory Mapping of Disabled Access. *Cartographic Perspectives* 41, 44–54.

Lefebvre, H. (1991). *The Production of Space*. Translates by D. Nicholson-Smith. Blackwell.

Liddell, H., Scott, R. and Jones, H. (1925). *The Online Liddell-Scott-Jones Greek-English Lexicon*, 9th edition. http://stephanus.tlg.uci.edu/lsj/#eid=2796 (accessed 17 April 2020).

Lynch, K. (2022). *Capitalism and Care: Why Affective Equality Matters for Social Justice*. Cambridge: Polity Press.

Marx, K. (1844). *Private Property and Communism (4), Third Manuscript; Economic and Philosophic Manuscripts of 1844*. Translated from German by G. Benton. www.marxists.org/archive/marx/works/1844/manuscripts/comm.htm (accessed 13 December 2023).

Menke, C. (2015). Andrea Butner's Aesthetics. *Parkett* 97, 33–45.

Mignolo, W. (2011). Modernity and Decoloniality. www.oxfordbibliographies.com/display/document/obo-9780199766581/obo-9780199766581-0017.xml#obo-9780199766581-0017-div2-0002 (accessed 13 December 2023).

Mignolo, W. (2012). Decolonial Aesthesis: From Singapore, To Cambridge, To Duke University. https://socialtextjournal.org/periscope_article/decolonial-aesthesis-from-singapore-to-cambridge-to-duke-university (accessed 13 December 2023).

Mignolo, W. and Vazquez, R. (2013). Decolonial AestheSis: Colonial Wounds/Decolonial Healings. https://socialtextjournal.org/periscope_article/decolonial-aesthesis-colonial-woundsdecolonial-healings (accessed 13 December 2023).

Moore, J. W. (2017). The Capitalocene, Part I: On the Nature and Origins of Our Ecological Crisis. *The Journal of Peasant Studies*. http://dx.doi.org/10.1080/03066150.2016.1235036 (accessed 13 December 2023).

Ngũgĩ wa Thiong'o. (1986). *Decolonising the Mind: The Politics of Language in African Literature*. Harare: Zimbabwe Publishing House.

Pearse, P. (1952). The Murder Machine. In P. Pearse, *Political Writings and Speeches*. Dublin: The Talbot Press.

Perec, G. (1999). *Species of Space and Other Pieces*. Translated from French by J. Sturrock. London: Penguin.

Pineau, E. (1994). Teaching Is Performance: Reconceptualizing a Problematic Metaphor. *American Educational Research Journal 31(1)*, 3–25.

Quayum, M. A. (2016). Education for Tomorrow: The Vision of Rabindranath Tagore. *Asian Studies Review* 2016, 1–16.

Quijano, A. (2008). Coloniality of Power, Eurocentrism, and Social Classification. In M. Moraña, E. Dussel, and C. A. Jáuregui (eds), *Coloniality at Large. Latin America and the Postcolonial Debate*, pp.181–224. Durham, NC: Duke University Press.

Rajkowska, J. (2020). I Hear Sounds Inside My Head. In B. Campkin and G. Duijzings (eds), *Engaged Urbanism: Cities & Methodologies*, pp. 200–204. London: I. B. Tauris.

Rancière, J. (2004). *The Politics of Aesthetics.* Translated from French by G. Rockhill. London: Continuum.

Rancière, J. (2008). Aesthetic Separation, Aesthetic Community: Scenes from the Aesthetic Regime of Art. *Art and Research* 2(1). www.artandresearch.org.uk/v2n1/ranciere.html (accessed 20 August 2020).

Rancière, J. (2010). *Dissensus: On Politics and Aesthetics.* Translated from French by S. Corcoran. London: Continuum.

Rice, L., Long, J., and Levenda, A. (2022). Against Climate Apartheid: Confronting the Persistent Legacies of Expendability for Climate Justice. *Environment and Planning E: Nature and Space* 5(2), 625–645. https://doi.org/10.1177/2514848621999286 (accessed 13 December 2020).

Rivera, O. (2022). *Andean Aesthetics and Anticolonial Resistance: A Cosmology of Unsociable Bodies*. London: Bloomsbury.

Sassen, S. (2014). *Expulsions, Brutality and Complexity in the Global Economy.* Cambridge, MA: Harvard University Press.

Schiller, F. (1793). On Grace and Dignity. Translated from the German by G. Gregory. *Friedrich Schiller: Poet of Freedom*, pp. 337–395. Washington, DC: Schiller Institute.

Schiller, F. (1794). *Aesthetical and Philosophical Essays.* www.gutenberg.org/files/6798/6798-h/6798-h.htm.

Shani, G. (2022). From 'Critical' Nationalism to 'Asia as Method': Tagore's Quest for a Moral Imaginary' and Its Implications for Post-Western International Relations. *Global Studies Quarterly* 2, 1–11.

Steyerl, H. (2007). The Empire of the Senses; Police as Art or the Crisis of Representation. http://eipcp.net/transversal/1007/steyerl/en (accessed 13 December 2023).

Steyerl, H. (2010). Politics of Art: Contemporary Art and the Transition to Post-Democracy. www.e-flux.com/journal/21/67696/politics-of-art-contemporary-art-and-the-transition-to-post-democracy (accessed 12 April 2020).

Suša, R., Andreotti, V., Stein, S., Ahenakew, C., Čajkova, T., Siwek, D. K., Cardoso, C., & Huni Kui, N. (2021). Global Citizenship Education and Sustainability Otherwise. In S. Saúde, M. Raposo, N. Pereira, and A. Rodrigues (eds), *Teaching and Learning Practices That Promote Sustainable Development and Active Citizenship*, pp. 1–23. IGI Global.

Transnational Decolonial Institute. (2011). Decolonial Aesthetics Manifesto (I). https://transnationaldecolonialinstitute.wordpress.com/decolonial-aesthetics (accessed 13 December 2023).

Tagore, R. (1961). The Problem of Education. In R. Tagore, *Towards Universal Man.* London: Asia Publishing House.

Tlostanova, M. (2022), Discordant Trajectories of the (Post-)Soviet (Post)Colonial Aesthetics. *Interventions* 24(7), 995–1010. https://doi.org/10.1080/1369801X.2022.2054003 (accessed 13 December 2023).

Williams, J.W. (2018). Freedom and Persuasion in the Attention Economy. D. Phil. Thesis, University of Oxford. https://ora.ox.ac.uk/objects/uuid:8065d7b0-4125-

4b1f-9218-8623380e03d7/download_file?file_format=application%2Fpdf&safe_file name=James%2BWilliams%2B-%2BDPhil%2BThesis%2B-%2BFINAL.pdf&type_of_work=Thesis (accessed 13 December 2023).
Yusoff, K. (2019). *A Billion Black Anthropocenes or None.* Minneapolis, MN: University of Minnesota Press.

10

AESTHETIC EXPERIENCES AND ARTISTIC EXPRESSIONS IN CLIMATE CHANGE EDUCATION

Leif Östman, Katrien Van Poeck and Ellen Vandenplas

Introduction

This chapter addresses the book's focus on exploring connections between arts and education in the context of the climate emergency from a pragmatist perspective. Säfström and Loughran (Introduction, this volume) emphasise that anthropogenic climate change symptomatically reveals the need for a shift away from top-down, command-and-control approaches to education. This implies not only striving for socialisation and qualification but also fostering students' ability to distinguish what Säfström and Loughran call their 'individuality and existential uniqueness', giving them the freedom and possibilities to develop their own responses (Van Poeck and Östman, 2020). We discuss how pragmatist theory enables us to investigate and understand the value of aesthetic experiences and artistic expressions for making this possible. We illustrate with empirical examples from a course on climate fiction how using arts (i.e. literature) in teaching practices can elicit aesthetic experiences and artistic expressions in relation to climate change and how teachers' careful didactic work can stimulate a critical and creative inquiry (Garrison et al., 2015) through which a new self and a new (view on/relation to the) world can co-emerge. We are especially interested in situations where the nexus of art and education can create 'teachable moments'. Such moments, Garrison (2010, p. xv) argues, occur 'when teachers and students engage in meaningful inquiry regarding some problematic situation involving themselves and the subject matter being taught'. These moments are precious and happen when participants are spontaneously deeply moved by a so-called 'poignant experience': a strongly felt disturbance that disrupts the one how has it in surprising, even shocking, ways.

DOI: 10.4324/9781003456636-11

> Such passionate experience dismantles expectations, frequently leading to a sudden redirection of thought, action, and inquiry. One experiences a desire to understand and/or express something new, exciting, and perhaps even a bit frightening. Care and concern arise as one becomes spontaneously engaged with unanticipated subjects of great interest.
>
> *(Östman, Garrison, and Van Poeck, 2024)*

It is in and through teachable moments that students get opportunities to develop their unique individuality, not as an innate property that autonomously unfolds but as something that takes shape and evolves though passionate and meaningful encounters with, for example, cultural artifacts (knowledge, books, films, etc.), values, and norms offered in teaching. Poignant experiences are omnipresent in the face of the devastating consequences of climate change and grasping precious teachable moments is crucial in pressing times where we need to creatively find new ways of living well with others—humans and non-humans—in the Anthropocene.

The aim of this chapter is to introduce a conceptual framework for studying the value of working with arts in climate change education and to illustrate and discuss what it allows us to grasp in empirical investigations. In what follows, we first discuss how pragmatist theory can be used to understand and empirically investigate the educative potential of using arts to elicit teachable moments and stimulate critical and creative inquiry in climate change teaching. We present: (i) a transactional theory of learning in which aesthetic encounters and experiences as drivers for inquiry take central stage; (ii) a conceptualisation of 'the artistic' that outlines four crucial conditions for artistic engagement; and (iii) a dramaturgical perspective on the art of teaching as a practice that involves 'scripting', 'staging', and 'performance'. Next, we apply the presented models to an empirical case: a university course on modern English literature focusing on climate fiction. We show how the teacher's didactic work elicited aesthetic experiences that resulted in varied outcomes and how some of them can be characterised as artistic expressions. We conclude the chapter by discussing how the presented framework can inform much-needed research on how climate change teaching can foster critical and creative inquiry through which new ways of being and living with others can emerge.

Learning through Inquiry Driven by Aesthetic Experiences

Transactional pragmatist philosophy inspired a model to understand learning (Östman et al., 2019) that makes it possible to investigate and explain the power of aesthetic encounters and experiences not only for the acquisition of knowledge and skills or the socialisation into certain values and norms, but also for people's development as unique persons. Figure 10.1 represents the main elements of the learning theory which we will briefly elaborate below (for an in-depth explanation see Östman et al., 2024).

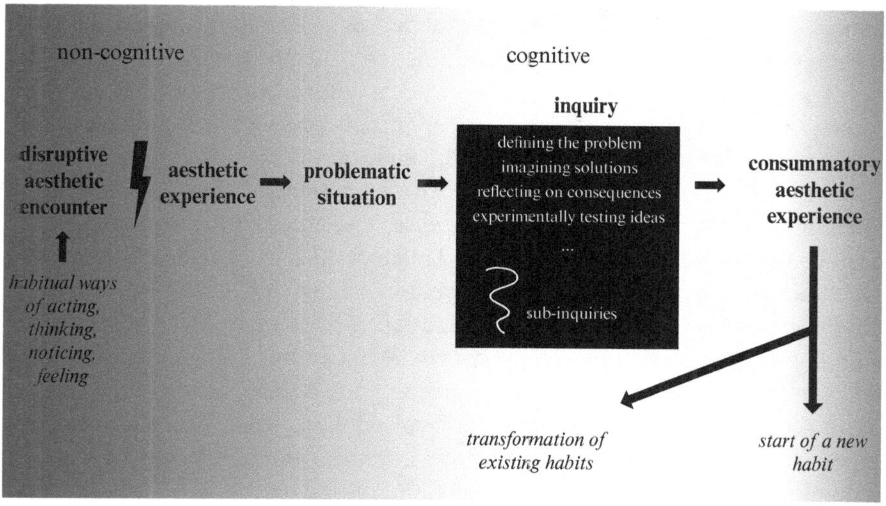

FIGURE 10.1 A schematic illustration of the learning trajectory.

Transactional learning theory conceives learning as taking place through inquiries driven by aesthetic encounters that disrupt our habitual ways of thinking, acting, seeing and feeling which gives rise to aesthetic experiences. Drawing inspiration from the radical empiricism of James (1976 [1912]), Dewey's transactional perspective on the entanglement of organisms and environments (Dewey and Bentley, 1989 [1949]) and his work on aesthetic experiences and artistic expressions (Dewey, 1987 [1934]) which Garrison (2010) elaborates on in his writings about the art of teaching, the starting point is that we are, from birth or even before, integrally part of the world. This anti-dualistic perspective of humans-already-embedded-in-the-world implies that an intellectual, cognitive way of relating to and being in the world is only one possible way besides, other, more fundamental ways of existing. James calls the latter 'pure experience' and Dewey calls it 'aesthetic experience'. Dewey (1981 [1925], pp. 377–378, emphasis in original) explains:

> *being* and *having* things in ways other than knowing them … exists, and are preconditions of reflection and knowledge. *Being* angry, stupid, wise, inquiring, having sugar … occur in dimensions incommensurable to knowing these things which we are and have and use, and which have and use us. Their existence is unique, and, strictly speaking, indescribable; they can *only be* and be *had*, and then be pointed to in reflection …. Their existence is … qualitative. All cognitive experience must start from and must terminate in being and having things in just such unique, irreparable and compelling ways.

Dewey's choice to call this type of experiences aesthetic is aligned with the meaning of the ancient Greek term *aisthetikos*, meaning sentient, sense, sensitivity, and perceptiveness (Andersson, Garrison, and Östman, 2018). Both James and Dewey saw this type of experience as pre-cognitive and pre-reflective and as constituting the basis of all cognition.

As Dewey emphasises in the quote above, existence (being and having) is incommensurable to knowing things and is, therefore, impossible to describe. This complicates the possibility to use language to intellectually communicate about aesthetic experiences because cognition cannot fully capture it. But if we are to explain intellectually what aesthetic experiences are, one can think of it as a bodily affect, noticed but not yet known. What is noticed, are changes in the body such as sudden muscle tension, arising goosebumps, increased heart rate, or an uncontrolled deep sigh. What is involved when being hit by an aesthetic experience, is bodily sensations through contacts, through being part of an encounter. The contact can be with anything in the surroundings (human and non-human), but also with our own memory. Or one can with the vocabulary of Todd (2023) say that one gets 'touched' by (an)other participant(s) in the encounter. One can either get touched as becoming connected through our senses (smell, taste, etc.), or in the sense that the connection is noticed. The latter is crucial since most affects remain unnoticed. In line with its pragmatist underpinning, the transactional theory of learning implies a non-dualist approach to body-mind and object-subject relationships. In that sense, Figure 10.1 may be somewhat misleading. Designed to represent the *trajectory* of a learning process, its two-dimensional form does not allow us to also represent the entanglement of aesthetic experiences and cognition. Figure 10.2 complements the representation by showing how pre-cognitive, pre-reflective situations provide the context 'for all inquiry' (Garrison, 2010, p. 86). While cognitive aspects are in the forefront

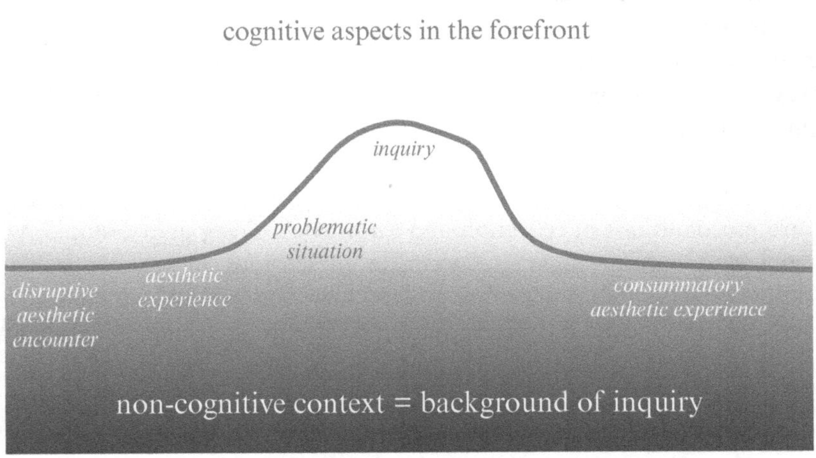

FIGURE 10.2 Non-cognitive background of cognitive inquiry.

during an inquiry, non-cognitive components like need, desire, sensory imagination, curiosity, and sympathy will always operate in the background.

Aesthetic experiences occur in transactional encounters with our surrounding. If resulting in aesthetic experiences, these encounters be called aesthetic encounters (Östman et al., 2024). The term transactional comes from Dewey and Bentley (1989 [1949]), who introduced it to contrast their approach to more mainstream perspectives of self-action and inter-action (Garrison et al., 2022). Trans-action assumes continuous, interdependent relationships between people and their (human and non-human) environment. The focus on how persons and their surroundings evolve constantly and mutually influence one another, significantly differs from the idea of self-action which suggests that entities, such as individuals, objects, or structures, possess intrinsic power to produce effects independently. It also differs from the concept of interaction which does acknowledge mutual influence between people and their environment but views phenomena as simply the sum of their separate components, linked through causal relationships between distinct entities. A transactional perspective, however, considers the situation as a whole, treating the components as part of an integrated system of coexisting organisms and environments and, thus, highlights the simultaneous, reciprocal changes in actor-environment relationships. From this perspective, aesthetic experiences, the meaning-making in an inquiry (see Figure 10.1) as well as artistic expression (see below) are created reciprocally and simultaneously in encounters of human and non-human participants.[1]

Two types of aesthetic experiences are crucial in a transactional understanding of learning processes (see Figure 10.1): those that constitute a disruptive initiation and those that constitute an integrative closure. The former occur when our habitual ways of acting, thinking, noticing, and feeling are disrupted and we cannot continue as usual in the activity we are involved in. Our habitual way of coordinating with our surroundings no longer function. We touched upon this above. The latter, 'unifying' aesthetic experience (see Figure 10.1) occurs when we have managed to restore a functional coordination with the surroundings, when we have overcome the disruption and can continue with the activity, be it in a different way than before. In-between both, the inquiry is situated. Let's further unpack these two moments and the process of inquiry.

Disruption of habits occurs through aesthetic encounters such as a car that does not stop at a pedestrian crossing, a sudden rain shower when sunbathing at the beach, an unexpected memory when reading a book that makes us stop reading. In these situations, an encounter with something in our world causes a disruption that is bodily noticed and constitutes an aesthetic experience that is pre-cognitive, pre-reflective and concerns the qualitative dimension of existence. James (1976 [1912], p. 46) describes it as 'a *that* which is not yet any definite *what, tho'* ready to be all sorts of whats'. An illustration is to just *be angry*, which has a quality that has no intellectual meaning until it is reflected upon (Dewey, 1984 [1930], p. 248). We have not yet made it intelligible, not defined it as a certain 'what' through inquiry. Although this quality is indescribable, we can characterise it as angry-ness. It is a quality that we can recognise bodily if someone says 'I am just angry, but I don't know who or what with'. They are then 'pointing to' a quality, in retrospect, by using rather general and vague descriptions of feelings of wonder, anger, curiosity, insecurity, scaredness, excitement, etc. Alexander (2016, p. 64) expresses this as follows: 'Our bodily involvement with the world pervades our unconscious experience and appears on the dim horizon of consciousness as a pervasive qualitative tonality or mood.' Only after further reflection, we can describe the reason for and the meaning of, for example, a state of insecurity as something that occurred when we heard about devastating consequences of climate change on the radio.

Qualities of a situation can be just enjoyed (e.g. the glimmering stars in a clear night or the touch of water enveloping our bodies when swimming) or, in other cases, push us forward. *Being* inspired, challenged, insecure, in doubt, curious, or perplexed can be compared to *being* hungry, thirsty, freezing, etc.: We are just brutally put in a state of a need that demands action. When the teacher asks a question on which the student does not have an answer—or vice versa—this need can be a state of being in doubt, an aesthetic experience that demands restorative action, which of course can be neglected, resisted, or postponed.

The non-cognitive, disturbing aesthetic encounter makes up an indeterminate situation which, if we decide to respond to it, turns into a problematic situation that initiates an inquiry. Dewey (1986 [1938], p. 100) explains:

> The indeterminate situation becomes problematic in the very process of being subjected to inquiry… There is nothing intellectual or cognitive in the existence of such situations, although they are the necessary condition of cognitive operations or inquiry. In themselves they are precognitive. The first result of evocation of inquiry is that the situation is taken, adjudged, to be problematic. To see that a situation requires inquiry is the initial step in inquiry.

The indeterminate situation functions as the context for the inquiry (see Figure 10.2). Dewey (1984 [1930], p. 248) describes its function as follows: it 'regulates pertinence or relevancy and force of every distinction and relation; it guides selection and rejection'. It is from the indeterminate situation that the 'data' (Dewey, 1984 [1929]) are taken that make up the raw material out of which, for example, an answer to a question is developed. It is important to emphasise 'taken': it involves selection influenced by earlier experiences. This taking involves intuition and selective attention and leads to discriminations: the taking of certain data and not others (Garrison, 2010, ch. 4). These data make up the subject matter for our imagination that produces ideas. Ideas have an important role in the subsequent steps of inquiry. They open up a *sense* of not yet realised possibilities in contrast with the conditions of the actual situation. This sense of what may be realised can function as an end-in-view for further reflection and experimentation through critical and creative inquiry (Garrison et al., 2015). It is the quality and intensity of the indeterminate situation that drives this process.

It is crucial to recognise that the process of inquiry is not as linear as commonly thought. Within the process of experimentation, reflection, etc. we constantly become part of encounters, some of which are disturbing and lead to sub-inquiries. Thus, an inquiry can consist of many sub-inquiries that crisscross and overlap with a complexity that can make us lose direction.

Restoring functional ways to coordinate with our surroundings through inquiry is ultimately necessary for our survival and well-being in everyday life (e.g. finding food for the day) or on a grander scale (e.g. combatting climate change). It can also be connected to an existential need and desire to flourish as a person through expressing subjectivity. And it can be connected to a demanding task given by the teacher, or the latter's demanding task to teach about climate change. In all challenges faced in everyday life, small or big, we are caught up in the rhythm of disruption and (potential) restoration through which persons and their environments continuously transform, reciprocally and simultaneously.

If an inquiry is successful and restoration reaches a closure, the unifying, consummatory form of aesthetic experience occurs: a direct experience of satisfaction or fulfilment. Like the disturbing aesthetic experience, this is an immediate, sentient experience, in this case not elicited by a disruption of coordination with the environment but by the formation of a new, restored coordination. Dewey (1987 [1934], p. 42) describes it as an experience that is 'demarcated in the general stream of experience from other experiences'. Everyone—artists, scientists, students, teachers, parents, athletes, etc.—have consummatory experiences and they can vary in quality and intensity.

Figure 10.1 illustrates how newness can come about in the learning trajectory. Aesthetic encounters and experiences make it impossible to just act and think as usual. Judging the indeterminate situation as problematic, forces us to reconsider and reshape habits but can also push the subject, the I to the front. If the problematic situation is taken as a personal concern, it creates an intrinsic motivation for the inquiry (see also below). New habits and new ways of relating to the world co-emerge with a new self. This co-evolution is crucial. The uniqueness of an individual subject is not an innate trait but something that is shaped through encounters and inquiry in which data are *taken* selectively, influenced by persons' earlier experiences as well as what is encountered in the environment. Also, it is important to emphasise that the integrative closure of an inquiry marked by a consummatory experience and (trans)formation of habits is only a temporary closure. In Dewey's (1987 [1934], pp. 62–63) terms: 'Fulfilling, consummating, are continuous functions, not mere ends, located at one place only'. Here, too, Figure 10.1 may be misleading as it could be interpreted to suggest that the unifying, consummatory experience marks a final and finite destiny of a linear learning process. Therefore, we complement it by Figure 10.3 to show how learning processes take shape circularly and continuously, whereby also newly (re)created habits can, again, become disrupted in subsequent aesthetic encounters.

Four Conditions for Artistic Engagement

We will draw on Dewey's work on the artistic (conceived much broader that 'museum arts': see below) which is driven by the question 'How is it that our everyday making of things grows into that form of making which is genuinely artistic?' (Dewey, 1987 [1934], p. 18). Dewey connects the artistic and the aesthetic to events made up of doing and undergoing through encounters with objects in the world:

> every experience is the result of interaction between a live creature and some aspect of the world in which he lives. A man does something … In consequence he undergoes, suffers something … The properties thus undergone determine further doing. … The process continues until a mutual adaptation of the self and the object emerges and that particular experience comes to a close.
>
> *(Dewey, 1987 [1934], p. 45)*

Four conditions are crucial for artistic engagement in the learning trajectory. First, artistic engagement involves love for the subject matter (the taken data) in focus for the inquiry: 'Craftmanship to be artistic in the final sense must be

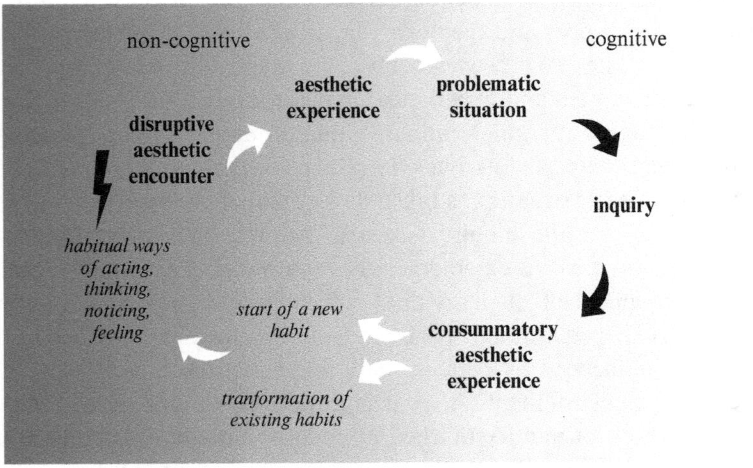

FIGURE 10.3 A non-linear representation of the learning trajectory.

'loving'; it must care deeply for the subject matter upon which skill is exercised' (Dewey, 1987 [1934], p. 54). This means to be open, to pay attention to, and to care for the indeterminate situation. This is done by setting up an emphatic communicative inquiry with the situation throughout the whole process of making. This requires—besides self-esteem, a supportive environment, etc.—a willingness to *surrender* to new aesthetic encounters and experiences, even in cases where one is already involved in an inquiry. In such cases love means to be open for sub-inquiries, that might lead to the abortion of the main inquiry giving precedence for a new one. Love includes giving up control for a moment, to let the situation disrupt, resist, and thereby give direction to subsequent doings and the creation of new subject matter. Woodcarving can serve as an example: one can just use one's force with the knife to carve out a format determined in detail in advance, or one can let the wood 'talk back' and let the form gradually take shape in the conversation.

Second, artistic expressions are personal. Aesthetic experiences stir something personal: 'Some occasion, be it what it may, stirs the personality … Then comes the need for expression' (Dewey, 1987 [1934], p. 78). It means, for example, that a highly personal idea is formed in the inquiry emerging from an aesthetic experience. It is not an idea that is part of a routine, neither is it forced through societal conventions and norms or purely elicited by the aesthetic encounter in itself. It is constituted through a junction between the new and the old, i.e. between the present, unique aesthetic encounter and the person's earlier experiences. The imagined idea functions as an end-in-view for expressing oneself in the subject matter. Personal ideas develop or even become substituted during an artistically driven inquiry in the undergoing of aesthetic encounters. The personal drive in the artistic inquiry is the reasons that there 'is an element of passion in all esthetic perception' (Dewey, 1987 [1934], p. 55).

Third, and relatedly, to 'be truly artistic, a work must also be esthetic—that is, framed for enjoyed receptive perception' (Dewey, 1987 [1934], p. 54). This means that, although the end-product anticipated can change and become more precise during the process, the process is driven by a desire for an aesthetical receptive perception of the end-product, i.e. to be liked by the producer and sometimes also by an imagined other, for example a teacher. The enjoyed receptive perception is also part of the artistic process. Every action, for example a carve by the knife in the wood, is done in the light of expressing the personal idea. As such, each carve, is judged aesthetically as negative or positive, as liked or dis-liked in relation to that idea. Without an idea of, for example something to carve, the abovementioned love for the subject would not have a direction and the whole series of criss-crossing and overlapping events of doing-and-undergoing would end without fulfilment. There would not be any artistic process, no events or products that would lend itself to be aesthetically enjoyed by the maker, just wood shavings.

Fourth, finally, artistic engagement involves spontaneity in the formation of ideas to be expressed. The personal idea is born in a communication with the subject matter. It is impossible to exactly know under which conditions such a birth will come about (e.g., in which conversation, when in the conversation) and even less what exactly the newly emerged idea will be. A spontaneously occurring idea for self-expression is suddenly wrung out of the conversation (Dewey, 1987 [1934]). Although it cannot be planned and guaranteed in advance, a person that is used to work artistically with, for example, wood can prepare the ground for creative ideas to be born and take care of them.

Imagine a task to make a spoon in wood, with a specific shape given in advance by the teacher. While working, one suddenly hits a hard piece (a knot) which was not visible from the start. The knife slides and makes a carve that disrupts the possibility to form the spoon in accordance with the plan. One can either stop and perhaps start all over with another piece of wood, or one can let the result of the disruption (the indeterminate situation) talk back. In the latter case of emphatically communicating with the situation, a new idea for how to design the spoon can suddenly come about, perhaps in a similar form as expected by the teacher. Here, the 'resistance' of the knot becomes an inspiration. The carver becomes excited and passionate. And the idea is spontaneous—it emerged in transaction with the wood—but also personal. The spontaneity and the personal are neither situated in the carver, nor in the wood. They are situated in the transactional event of undergoing-and-doing. In the continued inquiry for finishing up the spoon, the emphatic communication through the senses and movements become aesthetic, because it is driven by a passion for self-expression through the wood, making the senses highly alert, and because after each doing-undergoing a feeling of satisfaction is waiting to become realised and discharged. It is in such events in the inquiry that we often express our aesthetic perceptions and judgements of likes and dislikes with aesthetic words as cool, beautiful respectively shit and ugly.[2] If everything goes well, the end-product will be a piece of art for use and enjoyment in everyday life. Learning through the process will concern skills, knowledge, and norms as well as the growth as unique, sensing, and expressing persons with an individualised, personal communication with, and thereby connection to, the world.

Although personal gains are very high when becoming artistically engaged in a learning activity, we need to recognise that surrendering and still pushing a personal idea means putting oneself at risk. This also goes for a teacher.

The Art of Teaching

Teaching is an art. This is reflected in the meaning of the word 'didactic', derived from the ancient Greek group of words connected to *didaskein* (*didaktikos, didaskalia, didache*, etc.), which refers to 'teaching', 'showing something', but also 'playing out a drama' (Hoppman, 2007). Garrison (2010: xv) perceives teaching as 'a creative calling; it is poetry in the archaic Greek sense of poiesis, that is, calling something into existence'. Inviting students to become artistically engaged, to express themselves in the subject matter offers them opportunities to create something totally new for them: A new connection to the environment involving and implicating their own development as unique persons. Teaching as poetry implies that the teacher is artistically engaged and surrenders to create an emphatic conversation with students. When the artistic engagement of teacher and students meet, the teachable moment occurs. The desire to realise such moments that are not easy to achieve, is an attitude of the teacher that students will remember and highly appreciate. It offers the freedom to express oneself through the subject matter and to engage in creative making involving oneself in and with the world. But as in any teaching, it is in the end the students who decide whether or not to accept the offer.

A teacher can be artistic in both the planning and the performance of education practices. Transactional teaching theory (Östman, Van Poeck, and Öhman, 2019; Van Poeck and Östman, 2022) captures this in dramaturgical terms. The preparation involves 'scripting' and 'staging'. The former is about formulating purposes for the educative practice, clarifying the roles of teachers and students, explicating expectations regarding appropriate behaviour in this setting, etc. The latter consists of staging (i) a *scene* (i.e. determining what objects to focus attention on) and (ii) a *task* (i.e. determining how to engage with the object(s) in focus, what to do). Staging always involves planning encounters, aesthetic and others. Thus, the scene and the task will affect the doing and undergoing of the students.

In the process of planning (scripting and staging) the teacher imagines and anticipates the consequences of the emerging scripts, scenes, and tasks in terms of students' doing-and-undergoing (i.e. anticipating their receptive perception). For the teacher, this didactic imagination and anticipation can also bring about aesthetic encounters and, consequently, problematic situations initiating an inquiry with experimentation and reflection in the planning process (Schön, 1991). By virtually testing different arrangements, one can define problems and find potential solutions. Sometimes these aesthetic encounters and inquiries inspire the teacher deeply, making them to become artistically engaged in the planning, including aesthetic receptive perceptions and, hopefully, a consummatory aesthetic experience.

The teacher's planning may take into consideration how to enable students to become artistically engaged. For instance, formulating open instead of closed tasks, offering aesthetic encounters that can elicit inspiration, or just being prepared to support students if they would suddenly become artistically engaged. Not even the best prepared plan, however, can guarantee the occurrence of teachable moments. Careful didactical work of the teacher during the 'performance' (i.e. when implementing the plan) will always be vital. As mentioned above, artistic teaching requires an attitude of surrender to initiate emphatic conversation with the students. This may require that the scripted purpose of the lesson needs to be revised or postponed. In other cases, the purpose can persist but become widened to also give room for students' creative imaginations and their specific personal touch on the end product.

Aesthetic Experiences and Artistic Expressions in Climate Change Teaching

In what follows, we illustrate how the above elaborated conceptual framework can be empirically applied to investigate the educative potential of using arts to elicit teachable moments and stimulate critical and creative inquiry. Due to space limitations, we only illustrate the outcomes of the analyses and not provide a detailed explanation of the analytical methods and approach.[3]

We discuss three examples from teaching and learning practices in a university course addressing climate fiction in modern English literature. We observed how the scripting and staging elicited aesthetic experiences that resulted in varied outcomes—some of which can be characterised as artistic expressions. The teacher scripted the overall purpose of the course as engaging students in 'exploring how literature and culture more generally address the aesthetic, ethical, and existential challenges posed by climate change' (excerpt from syllabus). They are offered literary texts ('primary readings') and theoretical texts on, for example, the Anthropocene, climate justice, and climate anxiety ('secondary readings') and one of the assignments is to write response pieces focussing on how the assigned readings intersect and post these on a forum. In the syllabus, the teacher describes the staged scene (what to pay attention to?) and task (what to do?) as follows:

> Make sure to include close reading of the literary text and to support your analysis with textual evidence (citations from the texts). Your response should reflect on key concepts and ideas from the primary or secondary readings, highlight and analyse striking passages or motifs, assess the validity of arguments put forward, identify areas of agreement or disagreement, make connections with the class discussions, reflect on your reading experience, raise questions or topics for debate, and/or engage with posts from other students.

The students' writings were discussed in class. Another assignment at the end of the course was to write a reflection paper. This could be either a personal essay with an overall reflection on the course's content, or a creative storytelling of their own accompanied by an explanatory text.

Some of the students' writings in the reflection paper revealed how they were strongly touched by the teaching. One student describes the encounter with content of the first class as a shock. For this student, the disrupting aesthetic encounter was followed by an inquiry which resulted in an unexpected, new cognitive insight:

> The first class was already quite heavy for me. Hearing suddenly so much information that I was unfamiliar with was a bit shocking. I had actually never heard about the fact that we only have eight years left to do something about the earth warming up. I was flabbergasted because at that moment I realised that climate change was a direr problem than I anticipated.

Another student also describes a disruption that was surprising, here in the sense that she experienced a disruption of usually being able to handle triggering content easily. This aesthetic encounter, accompanied by an aesthetic experience of distress, was taken on as a personal challenge eliciting an inquiry into their mental boundaries:

> The trigger warning was not out of place for me, because this was indeed a topic that, to my surprise, caused me a lot of distress. Usually, I can handle a lot of triggering content considerably easily. It can of course disturb me a bit and I might think about it afterwards, but I never really felt as distressed as when I read about climate change. Sometimes I just had to put it down. Again, something I never experienced with other triggering content. So in that way, it was also kind of a challenge for myself to see if I had any mental boundaries and if I could overcome them.

The student's inquiry could be characterised as artistic. Relating the feelings of distress to the content of the course, and back to very different earlier experiences with triggering content, the student used the repeated aesthetic encounters and the aesthetic experience as a means for psychological self-inquiry. Engaging with the question whether she could overcome these boundaries could be seen as a way of trying to express herself in a contingent process, as a silent psychological artistic process. The artistic product, here, is the person one becomes, again and again.

The third example demonstrates a more classical artistic expression, where the media used is a song. As indicated above, students could opt for creative storytelling as an alternative to the standard reflection paper. The teacher describes the assignment and their experience of it as follows:

> I am in awe of the students in my climate fiction course. For the first time, I gave them the option to do a creative project as an alternative to the standard reflection paper. The assignment they received was to submit a piece of creative climate storytelling of their own, taking up any of the themes, tropes, or forms central to the literary texts we discussed over the course of the semester. To my pleasant surprise, many students jumped at the opportunity, creating short stories, (video) poems, songs, a text-adventure video game, a 1950s-style poster, and a manifesto against humanity. They evidently put a lot of thought, passion, and effort into these works, and in many cases the result just blew me away. … The reflection papers (there were still a good few) did not disappoint either, by the way. Here is my favourite line from them: 'It has essentially been a course about the experience of being alive in this time.'
>
> *(Social media post by the teacher)*

The teacher's scripting and staging made it possible for students to be creative with the material and concepts offered in the course. Describing this as 'a pleasant surprise' reflects a consummatory aesthetic experience of the teacher. As often for teachers, the main conditions for satisfaction concern students' engagement and results that 'blew them away' and elicited 'awe of the students'.

One of the students wrote a rap song and explained the background for it as follows:

> For me, music has always been a good way to express myself, and rap lyrics in particular allow for a certain kind of transparency that fit well with the message I had in mind. I wrote the lyrics from a perspective that embodies 'the West' during the Anthropocene, to confront our exploitative practices and the collective avoidance of their consequences. Throughout the course, the anthropogenic nature of climate change became an unavoidable fact. I realized that I had previously ignored the matter and that it was a privilege to do so. I tried to incorporate that realization and the acute worries and fears it evoked into this song. I also integrated concepts like deep time, climate injustice, and climate anxiety into the lyrics. As for the end of the song: I tried to end on a hopeful note, but seeing as I was left with so many insecurities, I ended with a question.

We observe here how the staged scene and task enabled the student's artistic engagement with the course's content. The student described an inquiry into concepts from the readings (e.g. Anthropocene, climate justice, climate anxiety) and the disturbing aesthetic experiences these evoked (e.g., acute worries and fears) resulting, here too, in new cognitive insights (e.g., realising that one had previously ignored the matter and was privileged to do so). Trying to incorporate these into the lyrics, the student also surrendered to the subject matter. While having an initial idea to end with a message of hope, the student allowed the felt insecurities to 'talk back' and this empathic communicative inquiry resulted in ending with a question—and a consummatory aesthetic experience for the student.

Conclusion

The above discussed empirical examples show how the pragmatist models of teaching and learning can be applied to deepen our understanding of the educative value of aesthetic experiences and artistic expressions in climate change teaching. We revealed how the teacher's didactic work, using artistic artifacts (climate fiction literature) and artistic practices (creative storytelling), resulted in aesthetic experiences, elicited by aesthetic encounters with the phenomenon of the climate crisis through the content of the course, which lead to outcomes as different as new cognitive insights, artistic transformation of the self, and artistic self-expression in creative writings. As illustrated most clearly in the third example where the artistic expression takes shape through traditional artistic media, the dramaturgy of the teaching practice facilitated artistic expression characterised by love for the subject matter, personal involvement, aesthetic receptive perception, and spontaneity. All three examples hold traces of teachable moments[4] and critical and creative inquiry into problematic situation related to climate change that deeply touched the participants, dismantled previously hold expectations, and redirected thought, action, and inquiry. The applied theoretical models allowed us to grasp how the dramaturgy of specific teaching and learning practices opened up a space for newness in a double sense: new selves and a new world (for them) in the context of climate crisis. These did not emerge in isolation, but interrelatedly, in and through aesthetic transactional encounters and aesthetic experiences. This happened though the offer of the teacher in terms of content, tasks, etc. The framework also revealed a crucial condition for this, i.e. the intrinsic motivation of students to express themselves artistically which emerged through aesthetic encounters. The value of creating such an educative space for newness is particularly pertinent in current times where we need to develop creative responses to socio-ecological crises in the Anthropocene.

Applying the presented conceptual framework in empirical research in diverse settings can deepen our insight into the didactical work needed to unlock the unique educative potential of aesthetic experiences and artistic expressions in climate change teaching. Here, it is important to emphasise that this should not be limited to teaching practices that involve artistic artefacts and practices. The presented models are inspired by pragmatist theory where the aesthetic and the artistic are naturalised, i.e. seen as central in any human activity, not only in arts. In 'Art as Experience', Dewey (1987 [1934]) explicitly rejects the idea to remit art to a separate realm (the museum) and sees it as integral to human experience and existence. As such, the conceptual framework is useful in all educational contexts while it remains interesting to conduct comparative analyses in order to explore whether and how artistic artefacts and practices may involve unique possibilities as often expected. More research about how aesthetic experiences and artistic expressions may occur across settings and subjects and under which conditions these can results in teachable moments and a transformation of persons and their environments is much needed. We hope that this chapter may inspire such studies.

Notes

1 Encounters can be perceived as what Massumi (2015) calls an event. In a transactional approach, an event consists of a continuous system of doing-and-undergoing (Dewey, 1987 [1934]). A disruptive aesthetic encounter can be seen as a specific form of event: being implicated in an encounter is an undergoing and the doing is that the activity one is involved in is stopped.

Almost immediately, the ending of the activity is another undergoing: an aesthetic experience of disruption.

2 In Wickman (2006) there are many examples coming from natural science teaching where students express these types of judgements with aesthetic expression of like and dislike.

3 As an example of the analytical approach used, see Van Poeck, Vandenplas, and Östman (2024) for a detailed explanation and illustration of how we applied transactional theories of teaching and learning in combination with Practical Epistemology Analysis.

4 Although the data-set did not allow us to observe these moments in the classroom, the students' reflective writings in retrospect reveal traces of them.

References

Alexander, T. (2016). Dewey's Philosophy of Art and Aesthetic Experience. *Artizein: Arts and Teaching Journal*, 2 (1), 59–67.

Andersson, J., Garrison, J., and Östman, L. (2018). *Empirical Philosophical Investigations in Education and Embodied Experience*. Cham, Switzerland: Palgrave.

Dewey, J. (1981 [1925]). Experience and Nature. In J. A. Boydston (ed.), *John Dewey: The Later Works, Volume 1*. Carbondale, IL: Southern Illinois University Press.

Dewey, J. (1984 [1929]). The Quest for Certainty. In J. A. Boydston (ed.), *John Dewey: The Later Works, Volume 4*. Carbondale, IL: Southern Illinois University Press.

Dewey, J. (1984 [1930]). Qualitative Thought. In J. A. Boydston (ed.), *John Dewey: The Later Works, Volume 5*. Carbondale, IL: Southern Illinois University Press.

Dewey, J. (1987 [1934]). Art as Experience. In J. A. Boydston (ed.), *John Dewey: The Later Works, Volume 10*. Carbondale, IL: Southern Illinois University Press.

Dewey, J. (1986 [1938]). Logic: The Theory of Inquiry. In J. A. Boydston (ed.), *John Dewey: The Later Works, Volume 12*. Carbondale, IL: Southern Illinois University Press.

Dewey, J. and Bentley, A. (1989). Knowing and the Known. In J. A. Boydston (ed.), *John Dewey: The Later Works, Volume 16*. Carbondale, IL: Southern Illinois University Press.

Garrison, J. (2010). *Dewey and Eros: Wisdom and Desire in the Art of Teaching*. Charlotte, NC: Information Age Publishing.

Garrison, J., Östman, L., and Håkansson, M. (2015). The Creative Use of Companion Values in Environmental Education and Education for Sustainable Development: Exploring the Educative Moment. *Environmental Education Research*, 21 (2), 183–204.

Garrison, J., Östman, L., and Öhman, J. (eds). (2022). *Deweyan Transactionalism in Education. Beyond Self-Action and Inter-Action*. New York: Bloomsbury Publishing.

Hoppman, S. (2007). Restrained Teaching: The Common Core of Didaktik. *European Educational Research Journal*, 6(2), 109–124.

James, W. (1976 [1912]). *Essays in Radical Empiricism*, Cambridge, MA: Harvard University Press.

Massumi, B. (2015). *Politics of Affect*. Cambridge: Polity Press.

Östman, L., Garrison, J., and Van Poeck, K. (2024). Poignant Experiences and the Nonteleological Teachable Moment. *Éducation & Didactique*, 18 (3), 91–110.

Östman, L., Van Poeck, K., and Öhman, J. (2019). A Transactional Theory on Sustainability Teaching: Teacher Moves. In K. Van Poeck, L. Östman, and J. Öhman (eds.), *Sustainable Development Teaching. Ethical and Political Challenges*, 140–152. New York: Routledge.

Schön, D. (1991). *The Reflective Practitioner.* London: Routledge.

Todd, S. (2023). *The Touch of the Present.* Albany, NY: State University of New York Press.

Wickman, P.-O. (2006). *Aesthetic Experience in Science Education: Learning and Meaning-Making as Situated Talk and Action.* London: Lawrence Erlbaum.

Van Poeck, K. and Östman, L. (2020). The Risk and Potentiality of Engaging with Sustainability Problems in Education—A Pragmatist Teaching Approach. *Journal of Philosophy of Education*, 54 (4), 1003–1018.

Van Poeck, K. and Östman, L. (2022). The Dramaturgy of Facilitating Learning Processes: A Transactional Theory and Analytical Approach. In J. Garrison, L. Östman, and J. Öhman (eds.) *Deweyan Transactionalism in Education. Beyond Self-Action and Inter-Action*, 123–136. New York: Bloomsbury Publishing.

Van Poeck, K., Vandenplas, E., and Östman, L. (2024). Teaching Action-Oriented Knowledge on Sustainability Issues. *Environmental Education Research*, 30 (3), 334–360.

CONCLUSION

Carl Anders Säfström and Glenn Loughran

As outlined in the previous chapters, the problem we are facing in this new post-climate landscape is that our fundamental uncertainty in post-climate times swallows all perspectives. There is nothing to see; there is no point from which we can stand on secure grounds to see what is in front of us, to understand what to do now. As Gert Biesta shows in Chapter 2, the idea that we 'can have a perspective' on the world rather than being in the world already does not solve anything but gives more of the same. We still imagine the ground, the foundation, to be the same as it has always been. Instead, as Biesta says, the world offers itself to us rather than ourselves, providing salvation to the world. As such, we are not in control but not enslaved since, within this givenness, the subject may appear. As such, emerging together with the world is always already there. Those moments of receiving the world instead of dominating the world are what Sharon Todd (Chapter 1) calls the black swan events, the educational moment and the moment of art, sensing our being so we can begin the world anew. As Todd says, the end-of-the-world scenarios we are constantly confronted with, rather than the end of the world as it is, are the end of a particular world projected onto our senses, where one world colonises the rest, one worldview dominating or striving for dominance.

Barbara Knežević (Chapter 4) suggests that since the forms of knowing being applied to life so far have led us to this situation of ecological crises, colonial forms of knowledge dominating all other forms, we can turn to sculptural knowing, a sensible form of knowing 'empathic, embodied, sensate, and full of regard for the agency of materials'. This is because such 'knowing' is already embodied and worldly and carries the seed for moving beyond the humanist self-centred world; it points towards a more-than-human world. It challenges us to step out of the map and instead touch the world to sense it. For Juliette Bertoldo (Chapter 7), we share this sensible world with those

DOI: 10.4324/9781003456636-12

beyond life. The dead touch us, speak to and teach us; the relation does not give away. We, therefore, carry a relational obligation to the dead, particularly those who are denied their death as well as their life, being in a state of migration, not recognised when living, so not in death. Denied their justice and killed twice over, they fall outside the map, or rather, they were never on the map, invisible. Here, mapping itself takes on another meaning; mapping those deaths brings them into relations they have been denied, bringing them into the world anew.

Paul O'Keeffe (Chapter 8) shows how migrants and refugees, ensembled in camps, in no man's lands, both literally and spatially, through art and education, are taking back their voices despite the violence of their situation. The violence of the colonial world, the systemic and physical violence the migrant, the refugee, and ourselves, as well as our students, face is not only theirs to suffer but ours to endure. Fiona Woods (Chapter 9) formulated a performative pedagogy to respond, bringing the affective to resist the destructive forces of our times and build confidence to counter the self-evident appearance of the social order, that there are other ways of being in the world, with the world. And that it is possible to, as Carl Anders Säfström says in Chapter 3, 'talk back from elsewhere'. Talking back, emerging from our 'plantness', and, as such, already being in the world as it *is*, outside the fundamental distinctions between man and nature imposed on us, imposed on our everyday life in which the polyvalence of languages deconstructs the drive of late capitalism to reduce everything to the map, which they already and with increasing sophisticated technology are in control over. For Tim Stott (Chapter 6), the mapping of the world, representing the world, takes form not only in the actual map but in art and as visualised in art history. Therefore, teaching art history today calls for a pedagogy that not only applies ecocritical art history but even, if necessary, also makes the way it is taught problematic by acknowledging the classroom itself as that place and space where beginning anew is possible—bringing art history into the conversation with pedagogy in the present. Leif Östman, Katrin Van Poeck, and Ellen Vandenplas (Chapter 10) offer a way of reading what is happening in a lecture hall in a classroom by applying a specific model formed through focusing on the aesthetic moment within an educational endeavour, and therefore, make problematic the instrumental tendencies built into education and teaching.

Finally, in Chapter 5, Glenn Loughran reflects on educational form through the lens of archipelagic thinking. Thinking about the form of education means considering how the spatial dynamics of environment and context shape educational processes and educational subjectivity. While agency in education often implies a shift in the teacher–student relationship, this chapter considers how the environment(s) act as agents within these educational forms.

Together, those chapters resist being trapped by more of the same: the idea that resistance demands us to be resilient to the shifting needs of late capitalism and that we should embrace ourselves for what is to come. Instead, we are already beyond such a point; the continuous value production of capitalism is driven by authoritarian nationalism of denial, clinging to the map, poisoning us all. The map is all that is left to the drowning humanity. Distanced and separated, literally creating an artificial intelligence grounded in the map itself as the only way forward. Therefore, resistance today means moving with others in a mixture of vulnerable lives beyond the map through a patchwork of overlapping sensibilities, recognised as such in their vulnerability and as beings open for the new to arrive. One way of doing this, to hack into the map, making the call, is what this book has done, to embody the art and education's eventful character, the closeness of life they both are expressions of, that energy that emerges in the moment, that autotelic moment, non-repeatable still adding value, making the moment inherently meaningful. No art, no education, is given meaning from outside of itself; instead, it gives birth to meaning, filling up the world from inside out. As such, it moves seamlessly between disciplinary expressions to embrace the life force it asks of us to be in the present, to embrace the world as it is. That moves us away from the domination of authoritarian capitalism to be in the world anew. The world is not, nor can be, a perfect whole represented by a masculine body of capitalism taking shape as the state of mind in eternity. Instead, as the teacher knows, wandering from city-state to city-state, from classroom to classroom, and as the artist knows, turning materiality into everyday life sensibilities, life is to embrace, to take the risk of living with people, animals, trees, mountains, and clouds in the *mixture*, in the mundane life, but as such, beautiful in its imperfections. To be fully present in the here and now is to be fully committed to what needs to be done in *this* world at precisely this point. The form of art and education taking shape throughout the chapters does not intend to mirror the world, but is about living the world and being in the world anew. Or, more radical, being *a* world anew.

INDEX

MIX
Papier | Fördert gute Waldnutzung
FSC® C083411

Kolibri 360 GmbH
Ferdinand-Jühlke-Straße 7
99095 Erfurt, Deutschland
produktsicherheit@kolibri360.de